NEWTON
AND
MODERN
PHYSICS

NEWTON AND MODERN PHYSICS

Peter Rowlands

University of Liverpool, UK

 **World Scientific**

NEW JERSEY · LONDON · SINGAPORE · BEIJING · SHANGHAI · HONG KONG · TAIPEI · CHENNAI · TOKYO

Published by

World Scientific Publishing Europe Ltd.

57 Shelton Street, Covent Garden, London WC2H 9HE

Head office: 5 Toh Tuck Link, Singapore 596224

USA office: 27 Warren Street, Suite 401-402, Hackensack, NJ 07601

Library of Congress Cataloging-in-Publication Data

Names: Rowlands, Peter, author.

Title: Newton and modern physics / by Peter Rowlands (University of Liverpool, UK).

Description: Singapore ; Hackensack, NJ : World Scientific, [2017] |
  Includes bibliographical references and index.

Identifiers: LCCN 2017002306| ISBN 9781786343291 (hc ; alk. paper) | ISBN 1786343290 (hc ; alk. paper) |
  ISBN 9781786343307 (pbk ; alk. paper) | ISBN 1786343304 (pbk ; alk. paper)

Subjects: LCSH: Physics--Methodology. | Newton, Isaac, 1642-1727.

Classification: LCC QC28 .R68 2017 | DDC 530--dc23

LC record available at https://lccn.loc.gov/2017002306

British Library Cataloguing-in-Publication Data

A catalogue record for this book is available from the British Library.

Desk Editors: Suraj Kumar/Jennifer Brough/Shi Ying Koe

Typeset by Stallion Press

Email: enquiries@stallionpress.com

Printed in Singapore

Preface

The Newton that is most familiar to us today is the consummate mathematician who, eschewing the hypothetical speculations prevalent among his contemporaries, gave us the method of defining fundamental mathematically-based laws that could be applied generically to a multitude of different physical situations. This is the Newton of the *Principia*, and the method he created and used in such a striking manner in that book has remained the ideal one for physicists ever since. The great physical theories of classical optics, thermodynamics, electromagnetism, special and general relativity, quantum mechanics and particle interactions were set up in exactly the same way and cover nearly the whole of currently accepted physics.

However, a very different Newton emerges from the *Opticks* and many of the texts that surround that work, where Newton was unable to find universal mathematically-based laws which could be used by his successors as a basis for further research. Here, if anything, with no sure mathematical method to guide him, Newton had to be even more original and inventive. The sheer power of the inductive approaches he used, by contrast with the mathematically deductive theorems that make up most of the *Principia*, can only be understood by comparing his ideas with those gradually worked out over the next 300 years leading to what we now recognise as modern physics.

The importance of this is that we have now reached a point in contemporary physics where the mathematically deductive approach pioneered in the *Principia* seems to be unable to suggest how to achieve the kind of inductive breakthrough that made the *Principia*'s fundamental mathematical laws discoverable in the first place. The less achieved programme of the *Opticks* and related works offers a better insight into Newton's more inductive mode of thinking because the lack of easy success forced Newton to be more creative in his attempts at explanation. The fact that his ideas so often prefigure the ones of later, more worked-out theories, suggest that his *method* of inductive thinking had special qualities which might be valuable for us to investigate today. This is the main objective of this book.

Though the bibliography and citations will show that I have taken full account of the discoveries and innovative explanations of many other scholars, the synthesis here is my own, and is, I believe, novel, along with many of the technical details. The work has been long-standing, and several aspects of it have appeared in earlier publications, sometimes reviewed by prominent Newtonian scholars. The philosophical views expressed are also my own and were published at a relatively early date. There are genuine novelties in the scientific part, both in interpretation and in technical detail, with results that are quantitative as well as qualitative. In general, however, I have made an effort to avoid both wholesale speculation and unnecessary technical detail.

One of the main purposes of this study has been to break down the fire walls that have been erected around particular scientific theories and to show how they only mean something when they are connected to a larger body of knowledge. As I see it, a true understanding of physics as it actually operates should see theories like special relativity, general relativity and quantum mechanics as having deeper connections as parts of a more coherent whole, connected with classical physics, rather than as being merely a succession of convenient models replacing each other as improved fits to the data — a picture that is more a result of scientists' promotional strategies than of the real nature of scientific development. Newton, as we will see, is surprisingly relevant to this process.

I have had the pleasure of meeting many of the Newtonian scholars whose works I have cited in this book. I am particularly indebted to Rupert Hall and Marie Boas Hall, who gave me encouragement when the project was at a very early stage together with permission to quote anything I wanted from their writings. Among contemporary scholars, I would like to single out Niccoló Guicciardini for reading through my drafts and making many valuable suggestions, but I am also grateful to Mike Houlden, Colin Pask, Mervyn Hobden, John Spencer and Stuart Leadstone for their comments. I have had stimulating discussions with many Newtonian scholars, including Rob Iliffe, William R. Newman, George E. Smith and Stephen Snobelen, and a considerable number of others cited in various parts of the book, though I wouldn't like to assume their automatic support for any of the opinions I have expressed. I would finally like to thank my wife Sydney for her constant and enthusiastic support and her help with the technical details of preparing the manuscript.

Peter Rowlands
Oliver Lodge Laboratory
University of Liverpool

About the Author

Dr. Peter Rowlands obtained his BSc and PhD degrees from the University of Manchester, UK. He then spent some time working in industry and further education. He became a Research Fellow at the Department of Physics, University of Liverpool in 1987, and still works there. He has also been elected as Honorary Governor of Harris Manchester College, University of Oxford, a post he has held since 1993. Dr. Rowlands has published around 200 research papers and 12 books. Some of his recent works, published by World Scientific, include *Zero to Infinity The Foundations of Physics* (2007), *The Foundations of Physical Law* (2014), and *How Schrödinger's Cat Escaped the Box* (2015). As a theoretical physicist, his research interests include, but are not limited to, foundations of physics, quantum mechanics, particle physics, and gravity. He has also done extensive work on the subjects of history and philosophy of science, and has published several books on these topics.

Contents

Chapter 1

Aspects of the Newtonian Methodology

1.1. The Modern Connection

Why do we want to write about Newton and *modern* physics? The immediate answer is that modern physics is not an achieved subject. It is still a work in progress. We have yet to uncover the true foundations, and there are many things about even the 'simplest' physical concepts, such as space and time, gravity and inertia, that continue to defy our comprehension. This means also that we haven't completely settled our account with the past. We don't know for sure that our present conceptions have completely replaced all the previous ones. The works of the past still exist in creative tension with those of the present. Though history of science is often written as though a succession of 'correct' ideas emerges to replace previously accepted 'false' ones in a series of 'revolutionary' developments, the history of physics isn't really like this. We have to only look at the history of wave and particle theories of light and material particles, the concept of the aether, ideas of absolute and relative motion, and even views of space and time held at different periods, to realise that the history of the subject is not purely one of monotonic development, and, if we insist on treating it this way, we will create barriers in our attempts at setting up an open-ended enquiry into a deeper understanding of the foundations.[1]

As a physicist interested, above all, in the foundations of the subject,[2] with many excursions also in the last 25 years into the separate disciplines of

---

[1]The case against the 'revolutionary' interpretation is made in Rowlands (1992, 1994). There are many other issues in physics that remain unresolved. For example, what is the most fundamental version of least action? What is the 'real' meaning of thermodynamics?
[2]See, especially, Rowlands (2007, 2014a, 2014b).

pure history of science, philosophy, and Newtonian studies,[3] I have come to
believe that Newton and the Newtonian method are of especial importance
to my own research programme, and are worthy of a treatment that cuts
across all rigid demarkations of genre as Newton frequently did himself.
So, while *Newton and Modern Physics* makes use of the vast literature on
Newton, written from historical, philosophical, scientific and mathematical
perspectives, the reader should not expect a pure example of any of these
approaches. The interdisciplinary perspective is deliberate because no other
is adequate for the purpose. Both contexts important to the history of
science, the contemporary or 'historical' one, and the more timeless, though
still not absolutely established, 'scientific' one, will be important, with
perhaps more emphasis on the latter. Newton is first and foremost an
important historical figure, but, as I hope to show, he is also still an active
participant in debates on fundamental questions. Nothing is included that
I don't believe to be historically true, but, where judgements are based on
conjecture, whether mine or that of another author, this is clearly stated.
The most significant of these is a method of thinking, which I believe
Newton used for many significant results. This is stated in Section 1.3, and
is based directly on my own experience. I am aware that it is not particularly
common, and the reader is free to disagree with my conjecture, but I believe
that several of Newton's writings show signs of its direct application, and
that, if accepted, it would explain many of the peculiarities of Newton's
thought.

So, why is Newton particularly important to a study which also involves
modern physics? The answer is that he comes right at the beginning of the
methodology that we now regularly employ without considering how strange
and counter-intuitive it is, where fundamental physical laws are generalised
statements about abstract concepts which have no direct relationship to the
things we observe in experiments, which are themselves understood only
as special cases operating under particular conditions. It would now be
unthinkable to employ any other method, but Newton's work is the first
that is recognisably modern in this sense, and there is a very large amount
of it in this vein in his writings. In addition, a study of his work, even in the
most obscure areas, suggests that Newton had a capacity to hit on the 'right'

---

[3]For examples of these respective genres, see Rowlands (1990, 1992, 1994, 2000, 2001,
2004, 2005, 2007, 2014a, 2014b); Hyland and Rowlands (2006); Rowlands and Attwood
(2006).

fundamental concepts, to such an extent that one cannot avoid noticing a remarkable tendency for modern developments to revert to ideas in some way close to his.

Now, classical optics, electromagnetic theory, thermodynamics, special and general relativity, quantum mechanics and the modern theory of matter are all clearly post-Newtonian theories — no one would claim that he anticipated them or even imagined that they would exist. But all of these theories incorporate ideas which are close to ones that he put forward and, where the ideas are close, Newton's are correct and correctly reasoned. The resemblances to modern developments are neither coincidental nor approximate. Also, though no one would claim that he anticipated the main ideas of his immediate successors, such as Maupertuis, Euler, Lagrange and Laplace, various authorities have already pointed out that there are ideas in his work which can be considered as special cases of these later developments. In addition, all the successful post-Newtonian theories have invariably been constructed on a Newtonian template. Even quantum mechanics, often presented as a significant break with the past, could equally be considered as an example of Newtonianism taken to a further extreme. The important thing from our perspective is that Newton's way of arriving at ideas which closely resemble aspects of later ones is often very different from those connected with the later developments, and yet it is still valid from its own point of view. This suggests that the resemblances are not due to Newton's 'prescience' but due to his use of concepts which are based on more fundamental truths than the ones we currently recognise. In effect, Newton gives us a 350-year baseline for tackling fundamental problems. His work gives us an indication of what kinds of ideas are significant in our own and what kinds of strategy we should use for reaching to even deeper levels of understanding beyond the superficial connections.

We should, therefore, be prepared to examine those aspects of modern physics where the foundational ideas are problematic, for example, the well-known incompatibility between general relativity and quantum mechanics, using whatever clues can be derived from a close study of developments across our extended time period. This means that we have to look at the history of physics in a way that goes beyond the linear revolutionary development model. As Newton's work tells us, there is often more than one approach to fundamental truths, while a closer examination based on the precise meaning of the relevant texts suggests that modern developments are not necessarily in conflict with older ideas, though they may extend them.[4] Of course, this is not the only reason for studying Newton's work. His creation of such a powerful method has significant intrinsic interest for both history and philosophy of science, while an extensive investigation of his writings in relation to a novel modern perspective can be expected to reveal numerous specific details not previously recorded.

1.2. Newton's Special Significance

A few years ago I was asked to write a counterfactual article on what would have happened if Isaac Newton had given up science in 1672, as he threatened, when his first published paper was subjected to a barrage of criticism from contemporaries attacking both his theories and his experimental methods.[5] Of course, I am a firm believer in the fact that historical events are post-determined in a uniquely ordered causal sequence, and so the counterfactual fails as physics as well as history. However, the exercise is still interesting in trying to isolate what is unique, or irreplaceable, about Newton's contribution to science. There is no doubt in my mind that Newton contributed something to science which was beyond the reach of any of his contemporaries, despite the outstanding talents that many of them clearly possessed. So, what was it, and what significance does it have for those of us who practise physics against a very different background several centuries later?

The key to Newton's significance seems to be that his mind and temperament were disposed to search for a generic solution at all times regardless of the starting point of the investigation, and to remain dissatisfied with any solution which included anything that failed to match this condition. This was true in every field in which he worked, in theology, for example, as much as in physics. In addition, he had the relatively uncommon ability to *recognise* the generic and to find ways in which to achieve it. Effectively, he saw that the key to achieving success in natural philosophy was to search for ideas which would simultaneously answer many problems instead of treating each one as a separate issue. This meant searching for the core ideas connecting areas that might initially seem unrelated, and, in the process, recognising

---

[4]This is especially true for general relativity, which (as we will show in the companion volume, *Newton and the Great World System*) is more of an *extension* of Newtonian gravity, rather than a revolutionary *replacement* of it by a completely new way of thinking. Even the use of geometry as a substitute for a force-type description has a precedent in Newtonian physics.
[5]Rowlands (2005).

and removing successive layers of accidental detail. Though this may be presented as a consciously worked-out process in retrospect, it is in fact almost impossible to do using standard 'scientific' thinking, for no deductive path will lead inexorably back to the origin from which the investigator's present knowledge has diverged.

Generic thinking requires the *simultaneous* use of abstraction, pattern recognition, mathematics, holistic vision, and uncompromising logic. Separately, these are all characteristic of Newton's thinking but their combination seems to reflect a type of thought process which is nothing like the conscious deductive development from prior starting positions, established from experiment or hypothesis, which is often claimed to be the classic scientific method. It is, in fact, very unusual to think in this generic way, and not always easy to justify the results to those who don't. Also, although the results, if accepted, may produce a major change in the scientific climate of the time, it doesn't necessarily follow that the methodology by which they were discovered will achieve equally widespread acceptance. In fact, this is what I believe happened to the work of Newton. The results were eventually accepted but the fundamental methodology was not, and it has taken a long effort up to the present day to begin to uncover it.

Newton's work is especially interesting today because it has very strong resonances with modern physics. Characteristic aspects of his work seem to have penetrated deep into the fabric of physics, even in areas he couldn't have known about. Many examples will be given in the chapters that follow, including more than a few that no one would expect to be associated with him, but, even in more familiar terms, it can be seen that his universal theory of gravity had to be supplemented by a worldview in which matter was mostly empty space, punctuated by hierarchical structures of tiny, almost point-like and virtually indestructible fundamental particles acting on each other with forces of attraction and repulsion. These forces were vastly stronger than gravity, and were assumed to be governed by laws diverging from gravity's simple inverse square relation. The electrical force was the most familiar and accounted for all the phenomena with which it is now associated, but there had also to be others, even stronger, at the very centre of matter, which had not yet been discovered. Nearly always, the more forward-looking ideas were backed up by calculations, both algebraic and numerical, which often produced results which are surprisingly compatible with those generated by modern theories. Because the emphasis of most historical scholarship is naturally focused on the context contemporary with the author, some of these ideas have never been studied in relation to the modern context. But such a study might well help us to discover those aspects of physics and physical theorising which are most significant at the fundamental level.

Because Newton's natural mode was to think generically, and to pursue his ideas as far as they could be pushed in this direction, his work in nearly all fields tends to stand up well when measured against modern views. This isn't to say that his views exactly coincide with modern ones or precisely anticipate them. Many things have intervened to make modern views special to their period. However, the Newtonian positions are usually *compatible* with modern ones within their limits and are not falsified by them. Claims to the contrary, as we will see in the analyses that follow, can usually be shown to be based on misunderstandings of either the Newtonian position, or the modern one, or both. This high rate of 'success' is nothing to do with some kind of magical 'prescience'. It is about choosing the correct line of attack from the beginning. Newton always looked for the kind of abstract generalisation which would deal simultaneously with many observations and generally avoided solutions which required specific model-building. He was not really interested in whether or not hypotheses based on models were 'true', because the real explanation, according to his way of thinking, would have some deeper underlying and non-specific cause.

Very significantly, also, Newton's success in science was largely driven by his need to create a metaphysics that satisfied his intellectual and spiritual needs. Until relatively recently, the science, once created, was considered an autonomous development, and the metaphysics nothing but an anachronism, which had only a passing historical interest.[6] Because Newton never wrote a separate treatise on metaphysics, he was disregarded in the history of that subject. However, it is now becoming increasingly clear that, while it was Descartes who set out a programme for a description of nature based upon first principles, it was Newton who managed to fulfil it, and he did so by creating a more powerful metaphysics than any of his contemporaries were able to achieve. Newton's scientific success was, to a large extent, a metaphysical success. All his major scientific achievements were logical consequences of deeply thought out positions on fundamental issues which were the province of metaphysics as much as of physics.[7]

---

[6]Spencer (2012) makes a strong case for the importance of metaphysics in the creation of physics, with special reference to the major physicists of the 20th century.

[7]In contradiction to the contextualising tendency found in some versions of intellectual history, Janiak's account of Newton's philosophy (2008) stresses its uniqueness, rightly in my opinion, for ideas which fit smoothly into the context of their time do not raise the kind of opposition which faced Newton during most of his life, and rarely have such an extraordinary impact.

Despite a long period of neglect, the connection is as significant for physics today as it was for physics in Newton's time. Physics is undoubtedly the most successful system ever devised for inquiry into the natural world, but it is not completely autonomous, and the foundations of physics cannot be understood purely within the context of physics itself. Without the use of a language that does not depend on arbitrary models and assumptions derived mainly from empirical evidence, physics will never discover the sources for its own foundations. The most surprising thing is that the only successful route so far discovered for penetrating to the deeper levels of physical truth was misunderstood as not being relevant to the subject it had created.

Bertrand Russell, in *Nightmares of Eminent Persons*, once remarked that there was a particularly painful department of Hell reserved for philosophers who had attempted to refute Hume's argument against induction[8]; maybe there is also a part reserved for physicists who have attempted to refute Newton and his 'inductive' methods![9] In fact, he resists refutation because of the fundamental, largely non-hypothetical way he thought. Though the Newtonian methodology was never fully established as the fundamental basis of subsequent physics, wherever physics has embraced it the results have been successful, and where it has strayed from the Newtonian line, problems have occurred. In fact, Newton's own physics has been the most fruitful source for future developments since his time. Though many claims have been made that aspects of it have been superseded, later positions have tended to revert to something closer to the Newtonian original. The continual reversion of modern physics to Newtonian or neo-Newtonian positions, which is much more widespread than has ever been acknowledged, is such an interesting and universal phenomenon that it deserves a name of its own — I call it the 'Newtonian attractor' — and it surely has its explanation in the generic nature of his characteristic mode of thought.

So, it is a particularly interesting exercise to look at Newton's work in relation to modern physical categories, such as quantum theory and thermodynamics, which he would not himself have used. In terms of these categories, his work makes sense because the underlying bases of these ideas are very similar to the ones he developed in his own work. If we

---

[8] Russell (1954, 32).

[9] In general, I will include the process that some philosophers and logicians call 'abduction' within the general label of 'induction', as earlier philosophers, including Newton, did, implying a method which is 'analytic' rather than 'synthetic', but without prejudice to the details of the method, which are not easily described.

put the positions Newton achieved in all aspects of physics of interest to later scientists to the test of detailed examination, we will also find out how far generic thinking of this kind can serve us at the present day. Since he achieved a qualitative way of abstract thinking that was almost mathematical, it is particularly important to put this alongside the directly mathematically-inspired results. The investigation could almost be said also to have a practical purpose. One thing that we have not yet achieved in modern physics is a fully unified theory. Now, Newton was undoubtedly the first scientist to realise that a truly unified theory of nature could be devised based on the fundamental principles of physics, and it may be that he still remains the best guide in our own contemporary search for a unified theory. Certainly, it would not be totally surprising if future positions had parallels within as yet unidentified aspects of Newtonian theory.[10]

In this kind of work, accumulated detail will provide the key information, and the detail needs to be as much physical as historical. We need to understand the concepts both as Newton used them and as we use them today, and we need to study Newton as much as possible in the words that he actually wrote. Paragraphs in Newton's work are often so condensed and loaded with ideas that we can read them numerous times without realising exactly what is being said. This is especially relevant in view of the fact that we are interested primarily in the way his mind worked, and only secondarily in the historical influence of his discoveries. This means that we are as concerned with drafts and abandoned documents, sometimes only recently made available, as much as in published texts with major influence such as the *Principia* and the *Opticks*. The historical influence of Newton's work is, of course, an extremely important study in itself, but its main relevance to our project is in allowing us to assess how far an individual who thought in a completely different way about fundamental questions was able to make an impact on his contemporaries and on later generations, and how much that we might think was potentially valuable had to be sacrificed in the process.

The historical *sequence* of Newton's texts is also important for his work almost always shows a drift in the direction of increasing abstraction and generalisation, and against the use of specific models and specific instances, and, where his starting points are non-rational sources such as biblical or alchemical texts, an increased effort at rationalisation. Like other thinkers in other eras, he started with what he had inherited from his predecessors, and

---

[10]Oddly enough, the recovery of the meaning of some of Newton's positions in later contexts seems to parallel his views on the retrospective meaning which could be recovered from biblical prophecies regarding historical events (see *Newton — Innovation and Controversy*).

then proceeded to adapt and modify in the way his own mode of thinking directed him. The ideas he began with emerged from many different sources with (in our terms) varying degrees of rationality, and with no consensus among contemporary scholars that scientific rationality had to replace all other forms of intellectual discourse.

The pattern that seems to emerge is that, wherever he began with an abstract generic picture, as, for example, when he was devising general mathematically-based rules for mechanics, his explanations stayed at that level. However, when he began with something more related to a 'physical' model, such as the actions of an aether in producing an effect like gravity, or a mechanical process as the explanation of an optical phenomenon, or something more esoteric such as the prescription for a process in the figurative language of the alchemists, or a cosmological event based on an interpretation of the bible, his tendency was to gradually remove the 'accidental' parts of the explanation to make it more abstract. Different interpretations of the same phenomenon could be worked on simultaneously by applying the abstracting process to ideas generated from different starting points. Ideally, they would eventually merge into a single coherent explanation. However, until that occurred, the most reliable was the most abstract, though the less abstract could still be worked on in the hope of achieving the final coherence.

Unless we understand this aspect of Newton's thought, we will be confused by the apparent inconsistencies of, say, a theory based on a rational approach to the mechanics of particles operating at the same time as a searching investigation of hundreds of alchemical texts written in the arcane and deliberately obscure language developed by alchemists over many centuries. Newton is not trying to impose a predetermined abstract pattern upon nature; he is trying to work back towards finding what patterns of this kind must exist for nature to behave in the way that it does, and he will start from the best evidence available either from his predecessors or from his own investigations. In this way, he will derive a totally abstract mathematical theory of gravity at the same time as putting the aethereal or 'physical' theories to the test of reasoning and experiment, and this will occur from the beginning because both methods are available, the first from his own theories of dynamics and Kepler's laws, and the second from the speculations of Descartes and other mechanistic philosophers. In the same way, he will copy out lengthy texts by alchemists based on extraordinary pseudomythological interpretations of what we must assume would have been heavily disguised real laboratory processes at the same time as carefully recording his own chemical experiments with an exactness unparalleled in the work of any predecessor, and then discussing them in published works in completely rational and distinctly chemical language. There is no rational/irrational split in Newton's work, only a varied rate at which he arrives (or sometimes fails to arrive) at the ultimately abstract conclusions. Newton is certainly complex, but not because he is ever truly irrational; rather, it is because he is working on several levels at once towards the final ideal.[11]

This must also affect our judgement of some of the most important historical questions concerning Newton's work. One, which is discussed in *Newton and the Great World System*, is the significance and meaning of the test of the Moon's gravity in relation to that of objects at the Earth's surface, which is represented by a manuscript of calculations from the 1660s which makes no mention of any 'physical' theorising with which they might be connected. The existence of such a manuscript seems to suggest that Newton was already well advanced in the abstract mode of theorising that subsequently became his signature method. There is no need to invoke any parallel theorising about Cartesian aether mechanisms for gravity. There is also no need to invoke any special influence from alchemy on the formation of the universal law of gravitation which was the ultimate result of his investigation. Universal and abstract mathematical laws were seemingly the tendency towards which his work was developing from a very early period. His particular cast of mind demanded it, and we should not imagine that, just because this way of thinking is unusual, he must have only developed it as the result of a great deal of prolonged reflection on esoteric subjects.

Before the middle of the 20th century much of Newton's work was not even published, let alone explored by scholars. After the tercentenary of his birth there was an explosion of scholarship. We now have the entire correspondence and a large number of biographies. Newton's life is pretty well known, though some aspects of it, such as the breakdown of 1693, remain enigmatic. The mathematics and many aspects of the mathematical physics have been relatively well served, with outstanding editions, translations, scholarly analyses and interpretations. Work is now expanding rapidly in areas never previously understood, such as the alchemy (or, more properly 'chymistry'), theology and prophecy. There are three major websites devoted to publishing the works previously unknown, especially in these areas, and

---

[11]Iliffe (2004) refers to Newton's 'great sensitivity to the disciplinary divisions' and describes how he 'simultaneously — and often radically — developed generically distinct concepts and ontologies that were appropriate to specific settings and locations.'

the Newton Project has published many of the recognisably 'scientific' works as well.[12]

I have made liberal use of all these sources and insights in my quest to understand how Newton actually thought. However, I believe I have also added many aspects to the interpretation, including some I first explored in three earlier books on the history of physics.[13] It seems to me that Newton's work as a conceptual thinker in physics has still to be fully elucidated, and this book is dedicated to uncovering all the areas in which he made an original contribution, so that we can obtain as complete a picture as possible. I have not hesitated to use insights from my work as a physicist to aid in this process, and I have used my own judgements rather than rely on clichéd interpretations which I believe are often far from the truth. The special importance of history of science is that it is about science as well as history, and the most successful results occur when we combine the expertise of the scientist with that of the historian.

The thing that has struck me most while writing this work is how remarkably correct nearly all of Newton's positions are, even after a century filled with 'revolutionary' developments such as relativity and quantum mechanics. Despite many statements to the contrary by both scientists and philosophers, hardly anything that Newton did is actually 'wrong', even in modern terms, and his work covers a ground that is far more extensive than most accounts would suggest. To be 'correct' on so many counts has, as we have said, nothing to do with 'prescience'. It suggests that Newton had both an archetypal 'system' and a method of working that we could still make use of if we could understand them better. The method, I believe, was the result of the particular cast of his mind and I have described the way I think it worked in Section 1.3. I don't think he would have considered it unusual because it was natural to him, but it is not common, even among the most gifted scientists, and I don't think it can be taught or acquired. It can, however, be understood, and that understanding would itself be of real value to the future development of science.

The system is more easily understood and it was based, in the first instance, on finding an abstract generic theory with the minimum specific assumptions. Clearly, Newton's tendency to do this was the result of the particular way he thought, but, once its characteristics have been identified, it can be applied on a more deliberate and systematic basis. Once the basic generic structure is understood, we can find more specific instances

---

[12]NP, NPA, NPC.
[13]Rowlands (1990, 1992, 1994).

for more restricted circumstances and do this iteratively to find increasingly exact explanations for particular phenomena. Prioritising generic features over specific ones at every level maximises the efficiency of the system's explanatory power, and makes it possible to apply it simultaneously in apparently diverse areas where the same patterns are repeated. It was, I think, the combination of this system and Newton's method that gave his work that particular character which enabled it to stand the test of time, and it is because of its success in this respect that it remains worthy of extensive and detailed attention from both scientists and historians.

Each one of the stories we will tell in the eight chapters has a different lesson for us. The present chapter indicates that there is a mode of thought outside of the accepted mathematical, theoretical, experimental, and hypothetical modes, and that some of the most significant aspects of Newton's work cannot be appreciated if we don't recognise this. At the same time, as we show in Chapter 2, Newton, the man, needs rescuing from failed attempts at psychoanalysis, which have coloured too many contemporary accounts. Following this, we introduce wave theory as the first technical subject. This is one that pervades physics, as Newton recognised, realising that it needed a thorough study from him, the first of its kind. The velocity of light, in Chapter 4, then provides a very complex story with many twists, while the chapter on mass–energy shows that Newtonian physics can still inform the present day in totally unexpected ways.

Many different developments related to quantum theory can be found in Newton's work; these are correct in themselves, though in nearly every sense counter-intuitive, awaiting their synthesis centuries later. The electric force, in Chapter 7, provides the great example of Newton's inductive method, but achieved too late for him to work out the details as he had for gravity. Following this, Newton's early realisation of a form of wave-particle duality shows his determination to hold on to fundamentally acquired truths whatever the apparent contradictions. The section on the 'unified field' shows a desperate thrust at a greater synthesis, which is still elusive today; but despite many denials, the version put forward by Newton is still the basic model for such attempts, with its necessary duality between non-local and local descriptions, and the use of a form of 'aether theory' which will recur in both general relativity and quantum field theory.

The central concern is with some of the less well-known aspects of Newton's work which resulted largely from his investigations in optics but which reach into areas that are now largely incorporated into quantum mechanics, relativity and electromagnetic theory. Newton also made significant inroads into areas which are now intrinsic components of astrophysics, cosmology, thermodynamics and the structure of matter, but these will be mainly covered in the companion volumes *Newton and the Great World System* and *Newton — Innovation and Controversy*.

Overall, the aim of this study is to show that Newton's work has a deep compatibility with areas of modern physics that he could never have imagined would exist, and this was because of the fundamentally generic way in which he pitched his thought, and the generalising power that his *recursive* method (described in Section 1.3) made possible. The generic structure that he created was so powerful that even now it provides a guide to future developments at the fundamental level, while compatibility with Newtonian thinking seems to provide the best route to resolving the major fault-lines that have prevented modern physics from becoming a unified subject.

1.3. The Power of Recursion

A brilliant pure mathematician, capable of creating an entirely new branch of mathematics and of developing others to unparalleled heights, an expert on the application of mathematical results to derive observed physical phenomena from fundamental laws in a way never previously attempted, an accurate and inventive experimenter with a practical streak, occasionally developing instruments and new technology, an inventor of hypotheses and speculative theories to suggest possibilities for experimental testing, an active research director supplying appropriate theoretical help and suggestions for experiments to more junior researchers, a trusted and capable public servant, using science and scientific method for the public good. These are the characteristics we might expect in the career of a quite exceptional mathematician and physicist, though it is perhaps unusual to be equally effective in all. They are all applicable to Isaac Newton and, as we know, there is a great deal besides — especially the private and extensive investigations into more arcane areas such as alchemy, prophecy, theology and ancient chronology — which suggests that, with Newton, we are dealing with a scientific thinker of almost unparalleled curiosity and inventiveness.

It would be difficult to find another individual who possessed anything like the same combination of skills in pure mathematics, mathematical and experimental physics, and the generation of new ideas by hypothesis, and this fact has been duly celebrated in a vast number of studies and biographies. Newton was, in the view of many, simply the greatest scientist in history, perhaps by a considerable margin. Yet, even in this picture, there is something missing. The procedures normally used to describe the scientific method do not exhaust Newton's ways of thinking, and, for this reason, a large amount of work that he did has never been fully understood or appreciated. A key additional methodology led to some of his most remarkable insights but has never been subjected to analysis.

Of course, this is historically interesting, but the interest isn't only historical. Though the key additional methodology is used only rarely by scientists, it is in fact very powerful, and could, if more widely practiced, make a significant difference to our understanding of physics at the fundamental level even today. Essentially, it is qualitative, though it usually also requires a deep understanding of mathematics. In fact, it frequently uses mathematics to gain important qualitative insights, and it can also be used directly in mathematical discovery — William Whiston, his protégé and successor at Cambridge, commented that Newton could see 'almost by Intuition, even without Demonstration.'[14] It effectively works by making the qualitative aspects quasi-mathematical, and it is one of Newton's most significant achievements that he produced something of a merger between mathematical and qualitative thinking by forcing the latter to the highest point of abstraction. Nevertheless, the qualitative aspect is the most significant. It is also inductive or analytical, rather than deductive or synthetic, though words such as 'induction' and 'abduction' completely fail to do it justice.[15]

Now, ever since Karl Popper developed a philosophy of science based on the so-called 'hypothetico-deductive method', it has been assumed by many that physics, and other forms of science, proceed by making hypotheses leading to deductive consequences, which are then tested by experiment. Some people have even denied that there is any such thing as an inductive process at all. However, though it is undoubtedly true that a great deal of science proceeds by hypothesis and deduction, some of the most significant developments with the most far-reaching consequences do not fit this pattern at all. Hypothesis-creation is essentially a very clumsy and laborious method which requires a great deal of iteration to produce meaningful results. Hardly anyone is actually 'good' at it — not even Newton — and it produces its effects only by steady accumulation and attrition. Newton did not like the method and repeatedly said so. What he preferred was much more like a recursive or holistic approach, which almost certainly uses an entirely different part of the brain. Here, the results are not 'hypotheses' in the ordinary sense because their self-evident validity becomes immediately apparent through such signifiers as intrinsic simplicity or generality of

---

[14]Whiston (1749, 39).

[15]'Reverse logic' might be closer. It suggests the effect but not the method.

application. Very often also the process does not require conscious effort — the results simply present themselves at the appropriate moment.

Obviously, we should ask, if such a method exists and has such amazing power, why don't we use it more widely and save ourselves from the trouble of devising endless hypotheses which will never exactly match the data we have in front of us until we have almost modified them out of existence? One answer is that a great deal of science is not concerned with the generalities that are best served by this method, but with particulars, where we have no option but to rely on hypotheses. Perhaps a more important one is that, with the way our brains appear to be hard-wired, it appears to be extremely difficult to do. In fact, it is not only difficult to do, but it seems to be also difficult to appreciate when the results are laid before us. Conscious thinking does not operate in this way and can't be used to appreciate it. However, it is not unusual for people to say that solutions to problems come to them when they have stopped thinking about them, the so-called eureka moment — Archimedes in his bath, Mendeleev dreaming of the Periodic Table, Kekulé on the Clapham omnibus, Bradley in his boat watching the pennant on the mast. So, some degree of such 'recursive' thinking might be available to many, even though it is rare for individuals to do it at the Newtonian level.

Newton needed all his methodologies to make the many breakthroughs that are associated with him, and the full account of his activities only makes complete sense when all these are considered together. However the method that we may variously call recursive or qualitative inductive or abstract analytical (though none of these gets really close to the way it operates) is of special interest because it leads to powerful syntheses of massive amounts of data, sometimes almost in an instant. To possess the ability of doing this on a large scale is an immense advantage in making scientific discoveries, but this comes with the disadvantage that hardly anyone else will understand the process.[16]

It is almost impossible to describe how it happens directly, but we can get some idea using analogies.[17] It seems to be like seeing the world as a mental stereogram. The mind makes an automatic *gestalt* switch, as with the Necker

---

[16]I should like to alert the reader at this point that the next two paragraphs are necessarily speculative, and, in true Newtonian style, should be regarded as separate from the rest of the discussion. Anyone wishing to avoid such speculation should proceed immediately to the paragraph beginning 'Examples in Newton's work'.

[17]The description here is based on comparing personal experience with an extensive study of Newton's writings. I don't propose to speculate on whether it comes from some particular region in the brain, but I am quite certain that it is an activity quite different from ordinary thinking as the two types of experience seem totally different.

cube. In a stereogram, a jumble of confused visual images suddenly reveals itself as disguising something completely recognisable — say, an elephant or an aeroplane — and this image, once seen, cannot be eradicated from the mind to return to the previous disorder. It is as though the mind gets to the point where it will no longer tolerate the chaos imposed upon it and self-organises into a simpler pattern through a phase transition. A problem cannot be solved by concentrating on one thing; there needs to be a buildup prior to success. The method appears to require a relaxed state, and works better with a great quantity of different materials, partially organised into patterns, than with a single set of information. It is almost easier to solve 10 problems this way than one, but it is not actually easy, and requires a lot of preparation. Thinking of this kind — parallel rather than serial — appears to improve with age, but the mind must be prepared by possessing a strong philosophy which can stand on its own. The results in effect, will be as good as the philosophy which produces them. More success comes with experience and with a wider field of knowledge outside science.

Remarkably, the discoverer knows instantly when an idea is successful. Certain patterns in the way it presents itself are automatically recognised by the brain as indicating correctness. These seem to include what amounts to an automatic check using something like Ockham's razor, seeking the minimum number of assumptions, though this is not itself a discovery method. Of course, it is possible the thinker may be deceived, but it is probable that those who are successful using the method are nearly always so, and their successes will eventually become obvious. One of the most striking things about Newton's writings is the strongly authoritative tone, derived from a feeling of apparent certainty, though he applies it only selectively. In fact, the air of certainty and authority comes only when he uses the inductive mode; when he is testing a hypothesis, he is considerably more tentative. *Pace* Frank Manuel, the author of *A Portrait of Isaac Newton*, this has nothing to do with Newton having a Messianic complex. The brain has simply exhibited closure. Another notable aspect of this way of thinking is that multiple connections are made in an instant, by a kind of avalanche process. Interestingly dream logic seems to be similar though it doesn't usually make actual sense. Like qualitative thinking in general, the ability at this kind of thinking appears to be rather uncommon, explaining why it is not well understood or appreciated. It is definitely not a product solely of intelligence, which is a necessary but not sufficient condition. Words like 'genius', 'insight' and 'intuition' are useless in describing it. 'Genius' suggests a degree of hyper-ability, but it doesn't specify what ability. Saying Ramanujan was a 'genius' doesn't explain how he actually did his mathematics, which was clearly by some process very different from that of the majority of other great mathematicians, who were also 'geniuses'.

Examples in Newton's work are legion — simply reading his texts gives the impression of it happening in real time, which is very probably what actually occurred. The early optical theory shows it on a small scale; equally early, but, more general, is the realisation of the third law of motion as a unifying principle in mechanics. The *Hypothesis of Light* of 1675 shows the process attempted at an early period on a large scale, but the lack of experience limits its success, and the *Hypothesis* is not so powerful an example of the analytical approach as several later works will be. The concepts haven't yet fallen into a form that has its own inevitability, and it is clear that Newton is only partially convinced himself. The final development of a totally abstract law of gravity shows it successful on a large scale in maturity. It took time, however, for Newton to develop the analytic powers to come to this rather startling conclusion, and to escape from decades of earlier indoctrination. Newton's very late treatment of the electric force is an even more powerful example, based on even greater experience, but this time he has no time or energy to pursue it. Various approaches in the direction of what we now know as the quantum theory show some local success, but more input will be needed to create a large-scale success. Many further examples will become apparent as we proceed through the chapters. Looked at in the light of the generality of this method in his thinking, Newton's story of the fall of an apple as the initial inspiration for universal gravitation is obviously true, the classic case of the *gestalt* switch (even though at this time only a partial one) occurring in a moment of relaxation. In view of the fact that Newton's manuscripts show this kind of thing occurring in greater or lesser degree on a more or less permanent basis, it is amazing that it was ever doubted.

The method, in Newton's case, involves a process of steadily removing accumulated detail by abstraction, and we can observe the discovery process as seemingly archetypal patterns reveal themselves behind the analyses. In this form good ideas can be assimilated from intrinsically non-scientific or semi-scientific areas, and it may be, through observing such archetypal patterns, that Newton was led to believe in ancient wisdom and to trust in dubious chronology and prophecy. Newton uses material from many such sources, including 'alchemy', as well as prophecy, chronology, and his belief in an ancient wisdom, somehow always reshaping it in a more scientific manner. However, if the patterns being investigated are not amenable to scientific thinking, the result can be bizarre, as in some of the historical analyses and the studies of prophecy. At the kind of level at which he frequently worked, simple ideas can repeat themselves so often that we can get the impression that the ideas must have existed earlier, as Newton believed must have been the case with the law of gravity. It can be established — as long as we realise the differences in meaning — that ideas developed at different periods may often have the same *basis*, though being frequently different in outcome. This is because simple ideas lie at the heart of nature, and this seems to be the source of the success with which Newton and a few others have had in absorbing non-scientific ideas like alchemy and theology, and even ancient history, into science.

Analogies frequently play a major part in the process, and Newton, at all times, worked by analogy and association to make experience gained in one field yield insights into another. In addition, the underlying structures always show through despite the technical difficulties. Though the moment when connections are finally made may be virtually instantaneous, it may have been preceded by years of constant thought, in which the concepts were worked on to bring them into line with observation. We see in Newton's work areas in which stylistic consistency is combined with continually changing perspectives, as, for example, in his ideas on the aether. In the early stages, an embryo form of the ideas may suggest future development in terms of powerful abstract categories, though the full development will require decades of further analysis.

Newton, in this mode, nearly always gives good leads for further work — apart from the usual calculus, gravity, optics, universal mechanics, abstract laws as the basic aspect of nature, a minimum number of separate principles, the use of space, time and mass, and no others, as fundamental parameters, and self-similarity at different scales, there are insights into areas that would now be termed thermodynamics, mass–energy, chemistry as high energy physics, particles as components of atoms, a hierarchical structure of matter with greater forces at the deepest levels, matter as mostly empty space, the interchangeability of matter and energy, wave-particle duality, forces deep within matter, electricity as the main force at the microlevel, a deeper level force to change the chemical nature of matter, electrical affinity in chemistry and the electrochemical series, and many others. Newton is always good when he uses this abstract analytical line of reasoning. He is much less good at coming up with hypotheses which require mechanical models, where the *gestalt* processes do not help at all. We could cite here his attempts to find corrections in the velocity of sound, and at finding mathematical laws of dispersion and double refraction. In fact, however, this is not a spectacularly successful area in scientific thinking as a whole, and it is difficult to understand how, in certain circles, the belief has been allowed to develop that this is the main process by which science advances.

Unless Newton's recursive method is understood, many of his results fall out of context and are seen as inexplicably prescient individual guesses instead of appearing as a systematic assault on the foundations of knowledge, which he felt unable to present in this way because he could not lay down a deductive path connecting them. Without understanding the method, we also fail to see the reason for his certainty that he was correct, despite concerns that the ideas would not be welcomed in the absence of any demonstration. As he well knew, this is nothing to do with hypothesising or model-building — this is a totally different brain register or way of thinking, providing 'other certain Truths'[18] — and it makes sense of his otherwise difficult-to-understand attack on hypotheses.

Once we do recognise the mode of thinking as relevant to Newton, however, the awesome power of the results is immediately manifest. The results are truly astonishing. Even the 'non-science' leads to remarkable conclusions. Though he didn't always resolve problems, Newton seems to be able to get under the skin of modern physics. It is difficult to read new pieces of his work without finding something of interest. Re-reading suggests new insights. New possibilities are continually being discovered in Newtonian ideas by modern workers, for example, the Newton fractal which emerges from an extension to Newton's iterative method of solving algebraic equations.[19] Newton's work is especially important today for anyone investigating the foundations of physics. His analyses often lead the reader to new fundamental meanings.

Among many other things, Newton was a highly inventive qualitative thinker, and the importance of qualitative thinking should never be minimised, for mathematical discoveries only make sense when they are put through the qualitative human 'transducer'. In qualitative thinking, he had a single method which he used to make a variety of imaginative responses. His philosophy was neither strictly causal nor mechanistic, and was far more profound than any that scientists have been allowed to use for the last 100 years. By necessity, he also created his own mathematics, whereas many 20th- and 21st-century physicists, for example, Einstein, have taken over theirs directly from predecessors in the 19th century. We can still read Newton for the benefit of gaining new insights which we hadn't previously suspected; for instance, on nonlinearity, on *vis inertiae* and conservation

---

[18]Q 31.

[19]Drexler *et al.* (1996) is one of many online discussions of this subject.

of energy, on force, momentum (especially in its use as a scalar quantity), mass, van der Waals-type interactions, and others too numerous to mention. Others may still lie hidden in his texts, not yet recognised. Laws like the third law of motion are immensely profound in their meaning, even in comparison perhaps with an easier-to-use and more popular principle like the conservation of energy.

However innovative, scientists have to use the language of their day in order to communicate with their scientific peers. As a result of this we often assume, in scientific history, that scientists' ideas are necessarily best seen in the context of those of their predecessors. This is an important aspect of scientific historiography, but, if used exclusively, it distorts the relevance of innovation, and implies an 'inevitability' in scientific developments which is quite often very far from the case. 'Causal' succession is a very different thing from inevitability. True innovators reshape available concepts in order to communicate to their contemporaries the significance of their new ones, but it does not necessarily follow that the fundamental basis of their ideas is that of the language which they adopt. This is especially true of Newton who has a habit of shading existing language into new meanings which make sense in his own world view. Some of the innovations of Newton, in fact, seem to be to a large extent independent of the background of the 17th century mechanistic philosophy with which they are frequently associated, and it may be that they have more meaning in a fundamental than in a contemporary context. Science itself gives us a means of approaching that fundamental context.

Newton, we may suspect, will always be the central figure in the history of science. It is difficult to deny that he was a quite extraordinary man by any standards — a prodigious thinker, enormously prolific, exceptionally versatile and resilient, and a great visionary. His style is now so completely stamped upon all things scientific that we cannot imagine how science would have developed without it. Powerful and authoritative, both as a man and as a scientist, he stands out even among the great scientists of the Western tradition. Newton more than anyone else is the source of our great confidence in the power of science, that this, and only this, is the way to ultimate truth. Practically every line he wrote on any scientific matter displays an incredibly clarity and certainty and a sure grasp of those ideas which we now regard as the fundamentals of scientific method. Newton's scientific work, as recognised today, incorporates outstanding achievement in four main areas: mathematics, mechanics, the system of the world, and optics. In each of these he created the scientific context in which we think and operate today. More significant than any of the particular discoveries he made in these fields is the fact that his system was universal. If we have to take any single individual as the originator of the quest for a unified theory of physics, and, by implication, the whole of knowledge, it has to be Newton. Though only a small part of the intellectual legacy of Newton was passed on to his successors to form the basis of what we now call 'Newtonian science', even this was sufficient to cause the most decisive advance ever made in the history of human thought.

1.4. The Newtonian Methodology

Many people, as we have seen, emphasise Newton's work in mathematics but he was at least as powerful as a purely qualitative thinker. One of his great feats was in bringing these modes of thinking together, so that they became almost versions of the same process. Contemporaries, however, who were convinced by the mathematics, were seldom prepared to accommodate the qualitative work which underpinned it, and much of it was never published, apparently with good reason — for Newton was unable to explain the method by which he achieved his insights. In public, he tended to stress the labour needed to accumulate the necessary information rather than the flashes of inspiration which made the labour worthwhile. In a letter to Bentley referring to his work on gravitation, Newton wrote: 'But if I have done ye publick any service this way 'tis due to nothing but industry & a patient thought.'[20] On another occasion he stated that he solved problems by constantly thinking about them.[21] The latter is certainly true. Putting together passages written perhaps 40 or 50 years apart gives, beneath the stylistic variations and fluctuating terminology, a sense of consistency in his approach. He tended to see very quickly the basic abstract pattern and to develop the categories to deal with it, but then to spend many years trying to put it into relevant words with relevant examples. In effect, he established categories of idea by some instinctive process, almost *a priori*, but with a strong empirical basis, and worked on them at intervals until he had matched them as closely as possible with observation. The categories would be simplified by elimination of superfluous information, and gradually sharpened by the assimilation of experimental data.

It sometimes seems as though the earlier embryo form of ideas would inevitably develop into the later more complex or more precise ones, as though the whole future development was actually contained within the

---

[20] 10 December 1692, *Corr.* III, 233.

[21] 'To one who had asked him on some occasion, by what means he had arrived at his discoveries, he replied, "By always thinking unto them;" ...', Biot (1833, 19).

more vague ideas on which it was founded. The future development would be conditioned by the more powerful abstract categories he had already created. This makes it difficult on occasions to track down the emergence of new ideas in his work and to discover the exact meaning of the terms he used at any particular time, but it does mean that later passages dovetail surprisingly smoothly with earlier ones, and that the later ideas do not normally contradict the earlier. Even when he changed his mind on basic causes, as in the case of capillarity after Hauksbee's experiments in the early 1700s, the description of effects is conceived in exactly the same kind of way and so seems hardly any different in the two versions.

With his powerful categories to help him, and his strong sense of the abstract nature of fundamental processes, he was able, as we have seen, to adapt or assimilate good ideas from many non-scientific or semi-scientific areas. We can see, for example, how he used his studies in alchemy and theology and channelled them into scientific theories, sharpening and redefining them with the mechanistic approach used by most of his contemporaries. If such sources of inspiration as theology, alchemy, prophecy, and even classical studies have sometimes been considered rather odd by rationalist critics, they become much less odd when we consider the thing that fundamental physics has become in modern times. It is often not really the sources of ideas that are important so much as how we modify them to fit in with a larger pattern; experience of areas of knowledge requiring an extension of the usual mental processes are as essential to creativity in science as they are to creativity in other fields, and perhaps to a greater extent. The origin of ideas becomes unimportant once they have been subjected to an abstracting process which removes most of their accumulated detail. That Newton believed that alchemy, for instance, was a source of fundamental truths was perfectly reasonable in the sense that it dealt with fundamental and recurrent issues using archetypal modes of thinking; and this is the way in which he used it, setting one author against another to find the common pattern which he believed must underlie the alchemists' figurative language. This search was justified in that it provided a set of abstract patterns for his own thought which would have been denied to a philosopher who could see value only in a mechanistic approach.

Setting side by side, without discrimination, ideas from separate periods of Newton's life, when his main concerns were quite different, allows us to perceive the archetypal patterns that lay behind his analyses of various phenomena. It also gives the impression of the endless searching so eloquently expressed in the letter to Nathaniel Hawes which Richard S. Westfall used in the title for his classic biography: '... he that is able to reason nimbly and judiciously about figure, force and motion, is never at rest till he gets over every rub.'[22] (The direct assumption of the reduction of scientific thought to the abstract concepts of figure, force and motion is striking here.) It allows us to see Newton in the very act of thinking. Telescoping the work of years and decades into the reading matter of less than hours simulates the very processes by which he arrived at answers to fundamental questions, the more so as his prose so often has an inquisitive tone, and not just in the Queries added to the *Opticks*. It provides a sort of sampling of thoughts arriving intuitively, as they must often have done, without the laborious reasoning and deductive processes which are so familiar from smoother accounts; some of the manuscript material must have been written, as we have said, just as he thought it out.

The juxtaposition of various sources shows how he thought all the way round a problem before developing his final answers, continually changing his perspective. His ideas on aether, for instance, are not a series of contradictory positions but a series of ever sharper definitions of the ultimate idea he sought. He was always prepared to accept an 'aether' as long as it was sufficiently immaterial to allow planets to move without resistance, and sufficiently material to be considered with the already existing category of matter, and also as long as it fitted in with his view of an active agent or agents, material or immaterial, responsible for the several forces.

Ideas from one period in his work frequently illuminate those of another. We can get a clearer notion of the limits between which any concept operated than if we only use ideas when either new or fully developed. Even taking passages out of context works in this regard. Ideas from one context were reapplied by him to others and he endlessly worked on them and renewed them. Few scientists have equalled his capacity for fruitful analogy. It is the sum total of all the statements that gives the true overall picture, and defines most precisely the nature of the individual notions he wished to express.

The concentration on the mathematical aspects of Newton's work tends to give a false impression of the nature of fundamental physics in general, and of Newton's unique contribution in particular, for there is a case for saying that he was not *primarily* a mathematical physicist. Empirical evidence was always the ultimate guiding principle for Newton, even when he abstracted beyond the particular cases investigated; he made mathematics continuous because nature was continuous. This explains a peculiarity of his preferred calculus, with respect to that of Leibniz, its association with the nature of

---

[22]25 May 1694, *Corr.* III, 357–366, 359–360.

time, for continuity in mathematics had to be explained with reference to a continuity in nature, and this is what time provided.

Given the initial principles, others could probably have done what he did in synthetic terms, that is in explaining individual phenomena, if over a considerably longer period, but the reduction to a few basic principles was a much more difficult enterprise and required expertise in apparently unrelated areas such as is seldom found in a single individual. We should perhaps seek the source of Newton's unique significance in the extraordinary brilliance of his qualitative thinking. If we continue to stress the mathematical side of Newton's achievement we are in danger of losing contact with Newton as an individual creator. A new system of mathematics, once created, may be made to appear, by some inevitable internal logic, to lead to mathematical laws of physics, and the thought processes which went into the creation of that far more revolutionary physics then disappear under a series of powerful, but somewhat impenetrable, mathematical expressions. Mathematical development is really part of the *synthetic* side of physics, the prediction and explanation of observed effects from given fundamental principles, not of the far more revolutionary analytic, which is concerned with the *discovery* of those principles. Analytical principles are not normally discovered as a direct result of mathematical approaches, but by a conceptual, and fundamentally *qualitative*, leap from some set of less well-organised data, which may be experimental, mathematical or conceptual in some other sense. Great analytical physicists need not be mathematicians at all, as the cases of Faraday, Rutherford, and even Einstein, will testify.

It is, indeed, a remarkable fact that synthetic developments are usually accepted without great difficulty, but analytical discoveries are seldom accepted without a prolonged struggle. Newton certainly astonished his contemporaries by his brilliant feats of synthesis. Even opponents like Leibniz and Johann Bernoulli commented favourably upon them. But his ideas in analysis, which were much more revolutionary, were completely beyond them. This is because they were not only derived by analysis, but were also *intrinsically abstract*. This meant that they could not be explained by the semi-synthetic models or mechanistic 'hypotheses' which have always been used to make analytically-derived truths more acceptable.

Cohen and Westfall list four Newtonian revolutions: in optics, mathematics, rational mechanics, and gravity.[23] Newton foresaw a fifth, in the investigation of electricity, but knew that he had neither the time nor

---

[23]Cohen and Westfall (1995), General Introduction, xi–xii.

energy to carry it through himself. As he himself would have imagined, this study has had a massive effect on the world, in technology as well as in the fundamentals of physics and chemistry. The chemical revolution of the later 18th, however, required a different mentality to the one supplied by Newton, though his work led to some chemical developments. There are many areas where he verged upon later ideas, some in mathematics (the calculus of variations, the Euler–Lagrange equations, the non-integrability of transcendental curves), many in physics, including moves towards ideas we now consider part of relativity, quantum theory, thermodynamics, and the modern theory of particle physics and the structure of matter. There are faint foreseeings of the forces that emerged with the discovery of radioactivity. There are also unresolved issues, leading to other aspects of thermodynamics, quantum mechanics and relativity.

History of science cannot be 'pure' history. It is writing about a privileged subject and it necessarily privileges science as being intrinsically significant. It doesn't make sense, therefore, to write history of science as though the science itself were unimportant. Newton belongs to today's mathematicians and physicists as much as to historians. The history, which involves the social conditions, the life and times, and relationship to other scientists and other pieces of science, is fascinating and tremendously important. The growth of science has changed the world in the last 400 years in more ways than any other human activity. So, we want to know how these changes happened, who were responsible for them, and what special circumstances led to subsequent developments. Newton is such an influential character that every aspect of his development is of interest, including the parts that, to modern ways of thinking, are antithetical to science. But the story is also of interest to working scientists, who are in constant dialogue with their great predecessors. To working scientists, Newton is as much part of the present as of the past, a part of the published literature of the subject with which we constantly engage.

Physics is a hard discipline, which needs to be learned anew over each generation. Part of our learning how to overcome difficulties is to see how our predecessors overcame theirs. In this sense, this book is my dialogue with Newton. In this context, we are not so much interested in the historical development of his thinking, but in the *character* of his thinking and its effectiveness as a method. So, here we are more interested in the connection with later work than we would be if adopting a purely historical point of view. We want to find out why his method works, where alternatives ones don't. We want to revisit the fundamentals and see where he breaks off and modern physics begins. For us, it is of special interest that the gulf between his ideas and ours is often surprisingly small. We need to look at where Newton got through, and where he stopped, or was stopped. We want to look at where he works at the edges, and where he knows there are open questions. We expect that the jagged edges may have something for our own science. Dualities, for example, appear everywhere where problems remain unresolved, both in Newtonian physics and our own, and there may be leads for us today in such things as they appear in Newton's writings. It is important for us to see where later physics picks up from where Newton left off, even when there is no direct knowledge of his work. This may or may not answer the same questions as investigating Newton's historical significance, but it is of equal importance.

Chapter 2

Newton the Man

2.1. The Early Career

Isaac Newton's life story is fascinating in itself and he has been the subject of numerous excellent biographies. The 20th century provided much great work on Newton, including several important accounts of his life and career, but a less acceptable aspect of 20th-century scholarship appeared in 1968 with Frank E. Manuel's brilliantly written but unfortunately massively defective *A Portrait of Isaac Newton*, which was based on Freudianism and its author's views on the monolithic nature of modern science. Its influence can be seen in all the biographical studies produced subsequently, despite its dubious scientific credibility. A lot of problems with the portrayal of Newton the man stem from this work and need to be eradicated. Its sensationalism and amateur psychoanalysis make it the inevitable basis for many more popular and less scholarly presentations of Newton's life and character. This is not to deny that, in other respects, Manuel was a great historical scholar, and introduced several aspects of Newton's life and work for the first time to serious historical study. The brief account that follows is intended to form an important context for the work of scientific discovery which will occupy the remaining chapters.

The story of Newton's early years is the stuff of legend. It is, however, undoubtedly remarkable that he was born on Christmas Day, in 1642, the year that Galileo died, and that he had no father, his own father, also named Isaac Newton, having died three months earlier; he was also premature and survived against the odds, a thing often taken in those superstitious times as a sign of the birth of a special person. Though Newton's father was an illiterate yeoman, he was nevertheless wealthy and owned a large farmhouse in Lincolnshire now called Woolsthorpe Manor. His mother, Hannah Ayscough, was from a more refined family, but one rather reduced in circumstances; her brother was a Cambridge MA. Despite his apparently humble origins, the wealth variously accumulated by his father and mother meant that Newton would never be poor. After the death of her husband, Hannah made provision for Isaac by marrying an elderly clergyman, Barnabas Smith, but when she went to live with him, it has been said, the boy felt abandoned and never completely forgave her.[1] He confessed in 1662 to 'Threatning my father and mother Smith to burne them and the house over them.'[2] There is no real evidence, however, of any lasting resentment of Newton towards either his mother or his stepfather, and there is no evidence either of any hostility to the half-siblings who were the products of their union.

There was certainly trouble about his education and it was apparently only on the initiative of a local schoolmaster, Henry Stokes, and his university-educated uncle, that he was, somewhat grudgingly, allowed to go to Cambridge in 1661. This followed a disastrous attempt at making the 'sober, silent, thinking lad' into a farmer; the servants at Woolsthorpe apparently 'rejoic'd at parting with him, declaring he was fit for nothing but the 'Versity'.[3] Even though he was potentially heir to a considerable estate, he was forced to go in the menial capacity of a subsizar. There was always a deep-rooted sense of insecurity about Newton which never left him, even when he became successful.

Newton's early university notebooks contain material from the pre-scribed university curriculum, which was still largely Aristotelian, but he was soon making notes from more advanced contemporary or near-contemporary authors, such as Descartes, Galileo and Robert Boyle. Réné Descartes had an enormous influence through his *Principia Philosophiae* (1644), which was a very significant pioneering investigation of natural philosophy from first principles. His work led to a school of similarly 'mechanistic' natural philosophers, and he became particularly significant to Newton. Essentially, Descartes believed that the world contained two different types of entity, minds, which could think, and bodies, which were characterised purely by extension. All the operations in nature were due to the fact that bodies were permanently in motion which could be transferred between them but never lost, but because bodies *constituted* extension, then there was no void space empty of matter because bodies and space could not be distinguished.

---

[1] This is supposition, as Newton was only three years old at the time of the marriage, and may have had virtually no memory of living previously with his mother.
[2] Confession, 1662, Westfall (1963). Newton cannot have been older than 10 at the time of this incident as Barnabas Smith died in 1653, and Hannah then returned to live with him at Woolsthorpe.
[3] Stukeley (1752/1936, 51).

The actions that we would now attribute to 'forces' with a transfer of motion were created by direct contact between adjacent bodies. The universe was an aethereal plenum filled with vortices which generated the observed motions of the planets.

Descartes defined certain rules that he believed were essential to the rational study of nature, and they included the use of mathematical descriptions of natural phenomena, but he did not suppose that the whole of nature could be worked out just from first principles. He also knew that he had to use results derived from experiment, and so he developed a method of explaining experimental results using 'mechanistic' or 'mechanical' models or hypotheses to make individual phenomena respond to his overall principles. Descartes' work offered a very attractive programme for other natural philosophers to follow, although many rejected the connection between body and extension, and opted for a version of atomism which had recently been revived with the recovery of classical atomistic texts. In the 17th century, nearly all natural philosophers adopted the Cartesian method or some hybrid between the Cartesian method and atomism. Despite its attractive features, however, and occasional local successes, Descartes' method was a compromise between several different approaches to nature, and, for all its appeal to first principles, it lacked logical consistency. It was capable of generating a ready explanation for virtually any physical phenomenon, and so was popular, but it failed the test of Ockham's razor (that is, of minimising assumptions), and so was vulnerable to anyone schooled in the more rigorous approaches of late scholastic philosophy.

The outcome turned out to be a long way from the original aims. This was not the kind of philosophy that would have lasting appeal for a man whose natural inclination ruled out all concept of compromise. So, though Newton, like nearly every other natural philosopher of the period, was at first seduced by Descartes' liberating vision, unlike most of his contemporaries, he quickly became perturbed by its many inconsistencies, and by its cavalier attitude to important aspects of theology, and it was not long before he began redefining Descartes' aims using a totally different metaphysics and methodology. He was also unable to accept the prevailing compromise between Cartesianism and the atomic theory that he had learned from other sources. If atoms and the void were true, then the whole of Descartes' reasoning based on the plenum and matter as extension had to be abandoned, especially if it also clashed with a theology based on an all-powerful and infinite deity, whose characteristics seemed most approachable through a concept like an infinitely-extended space devoid of bodies. In this sense, Newton's early negative reaction to Descartes would define most of his subsequent career, and it became the most important factor in the numerous clashes he would have with his contemporaries.

At Cambridge in about 1664, Newton's reading notes from other authors gradually developed into original entries on natural philosophy. Typically, uncompromising in his desire for the truth, he went so far in his researches as to try out some experiments which make us shudder today, sticking a bodkin, for example, behind his eyeball to observe the nature of the images produced, and staring at the reflection of the Sun for so long that he was forced to lie in a darkened room for several days to recover his sight. In an occurrence that was to be repeated on later occasions: 'He sate up so often long' to observe the comet of 1664 'that he found himself much disordered.'[4] At the same time he began to take up the study of mathematics based on the work of Descartes, and made considerable extensions to algebra and coordinate geometry. He was largely self-taught as a mathematician but he made astonishing progress within a few months. Cambridge is now considered one of the world's great centres for mathematics and physical science but in the 17th century things were different. It was then principally organised as a theological college to train young men for Anglican orders and mathematics was way down on the list of priorities. It was only when Newton was in his 3rd year that the first mathematics professor was appointed under the terms of the will of the benefactor, Sir Henry Lucas. Newton's interest was therefore somewhat unusual for Cambridge, and he was far from ideally situated to exploit his growing interest in physical science. The most notable thing about Newton's early years is his utter isolation from the scientific community. His approach to scientific work was formed quite independently of that of the newly-formed Royal Society and the whole context of his career is that of the outsider attempting to penetrate the inner circle of those who considered themselves to be the practitioners of the mainstream scientific tradition.

According to Richard S. Westfall, Newton's years at Trinity College coincided with the most disastrous period in the history of the College and of the University in general, a period hardly conducive to the kind of career he wanted to pursue. 'A philosopher in search of truth, he found himself among placemen in search of a place.'[5] But Newton was not entirely isolated. His career would never have got off the ground without help from a very remarkable man, the first Lucasian Professor, Isaac Barrow. Barrow is now a much underrated mathematician and scientist, but his achievements,

---

[4] John Conduitt, KCC, Keynes MS; Westfall (1980, 104).
[5] Westfall (1980, 190–191).

which included work in optics and important mathematical theorems that were precursors to the calculus, were, in fact, very considerable and it is clear that it was Barrow who at several vital moments interceded actively to help Newton to overcome the almost insuperable obstacles which lay between him and the security of a Cambridge fellowship.

Everyone knows the story of how, with the university closed as a result of plague in 1665, Newton spent a *marvellous year* in Lincolnshire making the remarkable discoveries with which his name will always be associated. In fact, the chronology of these years is not well established and the story should not be taken too literally, but it is clear that, at about this time, Newton:

(1) Discovered that the spectrum produced by white light was dispersed by a prism because white light was a composite formed of rays of different refractive index; though he realised that the range of possible colours was infinite, he eventually identified seven principal ones on an analogy with the musical scale — according to one interpretation (but see Section 3.3), this was the origin of the mysterious inclusion of 'indigo' as a principal colour.

(2) Obtained the most general description of the algebraic result now known as the binomial theorem, and used it in an investigation of infinite series.

(3) Discovered the general methods of the calculus, including the fundamental theorem relating the processes of differentiation and integration (or, as they were called at the time, the methods of tangents and quadratures).

(4) Developed an early form of his general dynamics, including early versions of his laws of motion, the formula mv^2/r for centrifugal force, and a new concept of force itself.

(5) Found that the force that made an object fall to the ground, and obey Galileo's laws of falling bodies, could be applied to explain the orbit of the Moon, and, inferentially, those of the planets round the Sun. By using Kepler's third law, that the squares of the planets' periodic times were proportional to the cubes of their mean distances from the Sun ($T^2 \propto r^3$), and mv^2/r, he found that the force would follow an inverse square law of attraction, though at this time he may have believed that it would also be balanced by a centrifugal law of repulsion. Newton's application of terrestrial gravity to the Moon's orbit was the occasion of the famous incident when he saw the apple fall in the garden at Woolsthorpe and he proceeded from his initial insight to do a numerical calculation, finding moderate agreement with his own theoretical prediction.

In extending his work on optics soon afterwards, Newton demonstrated the principle of periodicity for the first time in the experiment now known as Newton's rings. As an atomist from the beginning of his career, however, and inclined to extend this notion to the possibility of a particulate nature for light, Newton, over the years, developed a dualistic (wave-particle) theory which is remarkably similar to modern views on light. Extending his work on the spectrum Newton came to believe that mirrors should be introduced into telescopes to avoid the problems of chromatic aberration in lenses, and in 1668 he was the first to construct a fully working model of a reflecting telescope based on this principle. Newton had no doubt, however, that his outstanding contribution was the discovery of the different refrangibilities of the components of white light. He knew that this was a breakthrough discovery, something that very seldom came the way of an individual investigator, 'the oddest if not the most considerable detection which hath hitherto been made in the operations of Nature.'[6]

In 1669 Barrow, an ordained minister, resigned as Lucasian Professor in order to further his activities in the Church, and Newton was, at Barrow's insistence, made his successor. Barrow, who was one of the founding members of the Royal Society, had already been spreading the word about the brilliance of his protégé. In particular, news of Barrow's extraordinary find reached John Collins, a man who made a career of cultivating mathematical discovery in others more brilliant than himself. Collins received some of Newton's manuscripts and was astounded — in particular, he obtained from Barrow a treatise on the calculus, called *De analysi*, a work of outstanding brilliance and originality.[7] Newton, however, kept Collins at a distance, though he did meet him in London in 1669; he was unwilling to enter into the public domain. By 1671, however, he was unable to prevent himself becoming a minor celebrity when news of his reflecting telescope filtered through to London and the Society fellows. Newton was himself elected a fellow and, for once, he decided to test the waters on something which he considered much more important. In February 1672, he sent to the Society his first paper on

---

[6]Newton to Oldenburg, 18 January 1672, *Corr.* I, 82. Later on, he would be even more emphatic, describing it as a 'new theory never to be shaken' [draft for *Recensio libri* (1722), quoted Cohen (1999, 280)].

[7]*De analysi*, 1669, published 1711, *MP* II, 206–247. According to Carl Boyer (1968, 434), Newton's proof of Rule 1 at the end of this treatise was 'the first time in the history of mathematics that an area was found through the inverse of what we call differentiation', thus making Newton 'the effective inventor of the calculus.'

optics.[8] It was at first well received, but was soon subjected to criticism by the Royal Society's curator of experiments, Robert Hooke. Newton composed a furious reply, creating a breach with Hooke that was never totally healed; but Hooke's comments were followed by criticisms from many other scientists across Europe.

Controversy was the very last thing Newton wanted at this particular time, but he couldn't avoid being sucked into a seemingly never-ending dispute that also involved Christiaan Huygens, then Europe's premier physicist, the French natural philosopher Ignaz Pardies, and a group of English Jesuits based on the continent. Realising that he could hardly present himself in what might appear as a 'prophetic' role to a group of eminently reasonable men, Newton tried to convince them on their own terms, putting experimental results as the source of all his conclusions. When this failed, he resolved to concern himself 'no further about ye promotion of Philosophy' and withdrew into alchemical studies. According to Collins, writing to James Gregory late in 1675, Newton hadn't corresponded with him for a year 'not troubling as being intent upon Chimicall Studies and practices, and both he and Dr. Barrow beginning to think mathcall Speculations to grow at least nice and dry, if not somewhat barren.'[9] Both mathematics and optics were an unwelcome interruption to his current pursuits of alchemy and theology. The Newtonian method was thus nearly lost from scientific history and it was extremely fortunate that a second chance came some years later in respect of a totally different problem.

But scientific work was not the thing that at this time mostly concerned Newton. The Cambridge fellowship had been bought at a price. By 1675, he would either have to seek ordination in the Church of England or vacate the fellowship. A deeply religious man like Newton could not treat the ministry purely as a source for a comfortable income; he would have to immerse himself deeply in theology, and prepare for a major disruption in his scholarly pursuits. But, Newton was constitutionally unorthodox, and quite early in his studies, if not as early as 1675, he came up with something that threatened the whole of his new-found security, not to mention his already tormented conscience. During the course of any new study, he always filled a notebook with various headings under which he copied readings from the best authorities. So, in his main theological notebook, he started with the

---

[8]'A New Theory', *Philosophical Transactions*, No. 80, 3075–3087, 1672 (dated 6 February), *Corr.* I, 92–102.

[9]19 October 1675, quoted *MP* IV, xvi (1971). Barrow had, in 1673, returned to Cambridge as Master of Trinity.

most important concepts, taken directly from passages of scripture. His drift becomes evident right away. For example, *Deus Filius*, where he added the note: 'Therefore the Father is God of the Son [when the Son is considered] as God... Concerning the subordination of Christ see acts 2.33.36 etc.' And under *Deus Pater*: 'There is one God & one Mediator between God & Man ye Man Christ Jesus 1.Tim.2.5 The head of every man is Christ, & ye head of ye woman is ye man, & and the head of Christ is God. 1 Cor. II.3.'[10]

Almost immediately he turned to the other end of the notebook to write down his own thoughts on these readings. The headings express clearly what had from the beginning caught his interest: *De Trinitate, De Athanasio, De Arianis et Eunomianus et Macedonianis, De Haeristibus et Haereticis.*[11] He had become fascinated with Athanasius and the dreadful quarrel that had occurred in fourth-century Christianity over the Arian heresy. Arius, as the official storyline went, had been a heretic who had denied the Trinity and the status of Christ as God; Athanasius had saved the Church from this heresy, with the help of the Emperor Theodosius, but it had been a desperate, and often violent, struggle, best remembered as one of the darker episodes in the history of the Church. It would have been easier to have avoided bringing up such a painful and distasteful subject as these notebook headings clearly represented; after all the matter had been sorted out to everyone's satisfaction twelve centuries before; but Newton was not the sort of man to gloss over fine distinctions in the name of convenience. Not for him the quiet life; once he was on to something, he would not let go. He became fascinated by the whole story and the personality of Athanasius, and went to all the available early Church theologians to find the roots of the quarrel.

In his theology, at any rate, he began to see the worship of Christ as God as the most dreadful *idolatry* and, being the kind of man he was, he could not compromise. According to his best biographer, 'he set out at an early age to purge Christianity of irrationality, mystery, and superstition, and he never turned from that path.'[12] Of course, the worst of all idolaters were the Roman Catholics. After all, it was the Roman Church itself that had been responsible for the corruption from the beginning. 'If there be no transubstantiation never was pagan idolatry so bad as the Roman, as even Jesuits sometimes confess.'[13] And at this very moment here was a whole host of English Jesuits attacking his optical theory! Monasticism and

---

[10]Theological notebook, Keynes MS 2, c 1684–1690, Westfall (1980, 310).

[11]Westfall (1980, 312).

[12]Westfall (1980, 826).

[13]*Treatise on Revelation*, Yahuda MS 14, f. 9ᵛ, early 1670s, quoted Westfall (1980, 315).

hierarchical power had been the symbols of the corruption of Christianity, and monasticism owed much to Athanasius's patronage of St. Anthony. The Reformation had solved nothing, least of all, it seemed, in 17th century monastic Cambridge.

Newton identified himself with Arius; he desperately wanted to avoid controversy, but his unique style as a scientist made this inevitable. He was also totally on his own in theology; even his close Cambridge friend John Wickins was not let into his secret; and Barrow, the man to whom he owed everything, was an ardent Trinitarian, author of a *Defence of Holy Trinity*, who, immediately on his appointment as Master of Trinity College, declared that he would 'batter the atheists and then the Arians and Socinians.'[14] As a 'prophet' himself, Newton began to study the Bible as a prophecy of human history, not as a revelation into the life eternal.

Newton's theological and scientific studies also extended into areas of philosophy. It was possibly under the influence of the Cambridge Platonist philosopher, Henry More, that Newton gradually developed a revulsion from the philosophy of Descartes that even spread into his mathematics. At a later date he turned away from Descartes' analytical or algebraic style and back towards the synthetic geometry of Euclid which he had previously neglected; and he went so far in one mathematical MS of leaving a blank rather than write down Descartes' name![15] Newton's concern about the potential atheism in Descartes' separation of matter and spirit grew to alarm in a work entitled *De gravitatione et aequipondio fluidorum* which contains some of his first notions of absolute space and time, and some of his earlier treatments of dynamical subjects.[16]

With the day coming ever nearer when he would have to make the fateful decision to conform to the Church of England, and be deflected from his scholarly interests, Newton's career at Cambridge was saved at the last moment by a *deus ex machina*, when his protector, Isaac Barrow, now also a Royal chaplain, pulled strings with the easy-going Charles II to obtain a dispensation exempting the Lucasian Professor from compulsory holy orders; Newton must have made up some excuse about having no vocation for the ministry. It was as well that Barrow, committed and ardent

---

[14]'As to the first, it appears the doctor was prepared to batter the atheists and then the Arians and Socinians.' North (1890, II, 313).

[15]*MatShesos universalis*, July 1684, *MP* IV, 526–588; see Westfall (1980, 401).

[16]*Unp.*, 89–156. Feingold (2005, 26), has proposed that this work originated in a series of lectures given by Newton at Cambridge in 1671 against the mechanics of Descartes and the hydrostatics of Henry More, but that it was reworked about 10 years later into the document we now have.

Trinitarian as he was, never learnt about Newton's drift towards heterodoxy before his death in 1677.

For a time, the relief this caused made Newton freer and more communicative about his scientific work. In connection with the business of his dispensation, he travelled to London and called in at the Royal Society, where he found he was surprisingly well received, even by Hooke, who had successfully repeated some of his experiments. He felt encouraged enough to send a major statement of many of his views relating to optics, *An Hypothesis of Light*, to the Society in December 1675, along with an extensive *Discourse of Observations*, which would form the basis of his later *Opticks*. It was the other forbidden fruit, alchemy, which lay behind the poetry of his vision in *An Hypothesis* of a universal aether responsible for everything in nature:

> Perhaps the whole frame of Nature may be nothing but various contextures of some certaine aethereall Spirits or vapours condens'd as it were by praecipitation . . . and after condensation wrought into various forms, at first by the immediate hand of the Creator, and ever since by the power of Nature, wch by vertue of the command Increase & Multiply, became a complete Imitator of the copies sett her by the Protoplast.[17]

But his theology can be detected here too. Though Hooke took exception to some apparent criticisms in the work, they managed to make it up, with Newton saying that he had seen further only by 'Standing on the shoulders of Giants.'[18]

In the meantime, a new figure had appeared on the scene. This was Gottfried Wilhelm Leibniz, one of the truly universal geniuses of the age, profound philosopher and mathematician, with innovations to his credit in such areas as binary numbers, determinants, calculating machines and mathematical logic. Leibniz had visited England during the early part of 1673, as part of a diplomatic mission. Neither the diplomatic mission nor Leibniz's personal scientific one was an unalloyed success. He had visited the Royal Society, who would elect him a Fellow on 19 April, but, while there, he had a confrontation with none other than Robert Hooke, who savagely criticised the prototype calculating machine he had constructed, and then, a month later, in characteristic style, produced one of his own, which Leibniz thought was a direct copy. Leibniz also, on another occasion, met the mathematician John Pell who informed him (correctly, as it turned out) that the results he had been claiming as his own on series and interpolation using successive differences had already been published in Lyons in 1670 in a

---

[17] *Hyp.*, 1675, *Corr.* I, 361–386, Newton to Oldenburg, 26 January 1676, *Corr.* I, 364, 414.
[18] Newton to Hooke, 5 February 1676, *Corr.* I, 416. See Westfall (1980, 272–274).

book written by Gabriel Mouton.[19] This, however, was right at the beginning of Leibniz's career in mathematics. He soon developed a remarkable facility in the subject, his progress being in many ways similar to Newton's early development a decade earlier, though without the same degree of sheer natural facility that Newton seemed to show when starting any new subject.

By April 1675, Leibniz had received some reports of Newton's achievements in mathematics, especially in his method of infinite series. This was in a report by John Collins which Henry Oldenburg, the Royal Society's Foreign Secretary, transmitted to him, but there had been an even earlier report, 2 years previously, which had given an outline of Newton's discoveries extending beyond his work in series. In addition, there was a letter from Oldenburg of 8 December 1674, which outlined the successes of both Newton and James Gregory in finding the length of a curved line, given an ordinate, and also the area, the 'centre of gravity, the solid of revolution', and 'the converse of these', as well as methods for computing logarithmic sines, tangents and secants. It was Leibniz's reply to this letter which had led to Collins's report, transmitted by Oldenburg on 12 April 1675.[20]

Then, late in the same year, according to the consensus view which has developed over the last two centuries, Leibniz discovered essentially the same principles of the calculus that Newton had discovered 10 years before, though he expressed it in a slightly different form, with a significant emphasis on symbolism and notation which was absent from Newton's work. In 1676, he wrote to Newton in flattering terms receiving from the latter two very lengthy replies, in one of which was a quite untranslatable anagram stating the basis of Newton's new method.[21] Thoroughly turned off by the protracted

---

[19]Brown (2012, 111–113).

[20]Oldenburg (1964–1986), IX, 563–567 (9 April 1673); XI, 139–140 (Latin), 141 (translation) (8 December 1674); 253–262 (Collins's English draft), 253–262 (Oldenburg's Latin (12 April 1675), 265–273); Hall (1992), 253–254. Brian E. Blank (2009, 603) observes that Leibniz's reply to the 1675 letter was 'not candid', in a way that seems to have become characteristic with him, implying that he had obtained his own results but not being prepared to produce the evidence.

[21]Leibniz to Oldenburg, 2 May 1676, *Corr.* II, 3–4; *Epistola prior*, 13 June 1676, *Corr.* II, 20–47; *Epistola posterior*, 24 October 1676, *Corr.* II, 110–161. Whiteside calls Newton's letters 'majestic' (*MW*, xi) and says that the second was written in a tone 'of friendly helpfulness' (*MP* IV, 671–673). Westfall's detection of an 'unpleasant paranoia' (1980, 267) seems to be coloured by hindsight. Hall thought the letters 'no mean act of generosity' and describes the first as 'one of the more generous productions of Newton's life' (1992, 156, 155). Richard C. Brown detects only politeness and friendliness (2012, 217). While Leibniz's correspondence seems to imply a degree of 'economy with the truth', Newton's ostensible politeness, according to Blank, shows him to be a master of the art of damning with faint praise (2009, 605).

controversy over his optics, Newton specifically required Oldenburg and Collins not to publish his mathematical work, but he made sure of protecting his priority. In October of the same year, before he had communicated his own discoveries, Leibniz came to England and met Collins. Among the manuscripts he studied, quite unknown to Newton, was Newton's *De analysi*, and also his letter to Collins of 10 December 1672, where he claimed that he had a general method of tangents, which, unlike those of Barrow and de Sluse and other early workers, extended to equations containing 'surd quantities', and could be used for 'mechanical' (or transcendental) as well as 'geometrical' curves, and for obtaining the 'crookedness, areas, lengths' and 'centres of gravity of curves.'[22] Leibniz found the basis of this tangent letter incorporated in the *Historiola*, a draft work on the history of British mathematics which Collins had started to prepare after the correspondence of the preceding years, though this had more work by Gregory than by Newton.

Newton's second letter, the *Epistola posterior*, did not reach Leibniz until June 1677. He then immediately wrote to Oldenburg with full details of his differential calculus and its fundamental algorithms, at the same time requesting further correspondence with Newton. Though Newton had withheld some of his most advanced methods, Leibniz's letter shows that he had either seen or suspected that Newton had developed procedures similar to his own. He wrote a second letter in July, but Oldenburg wrote back in August saying that Newton had other preoccupations.[23] Newton himself did not receive the first letter until 30 August. Four days later, Oldenburg died suddenly, extinguishing whatever hope there might have been that a fruitful cooperation between the two mathematicians could be established.

Still involved in the long and gruelling battle over his optics, and desperate to avoid further controversy, Newton refused to allow his innovative mathematical work to be printed, although he did allow a limited circulation in manuscript form. According to Guicciardini, Newton may have felt that his claim that mathematical certainty was the basis of his optics was vulnerable to the accusation that his novel mathematics, like that of the heavily-criticised John Wallis, was merely heuristic and not based, as he was now beginning to believe, on the certainties provided by the geometrical methods of the ancients.[24] The sequel to this was probably

---

[22] *Corr.* I, 247–252, 247–248.

[23] Leibniz to Oldenburg, 11 June 1677 *Corr.* II, 212–219, 12 July 1677, *Corr.* II, 231–232; Oldenburg to Leibniz, 9 August 1677, *Corr.* II, 235.

[24] Guicciardini (2009, 331, 344). Westfall (1980, 267), says that, in not printing his work and in not being explicit in his correspondence about his most advanced methods, Newton 'forced Leibniz to share' the discovery of the calculus with him, but he did not force him to claim sole authorship.

the most unpleasant quarrel in scientific history, eclipsing by far the long controversy over Newton's optical work, though, in principle, it was much less important.

2.2. The Later Career

Newton's isolation was interrupted only when the Royal Society decided that he alone had the mathematical resources to tackle the unsolved problem of the force law for planetary orbits. News of Newton's secret progress in higher mathematics had reached the Royal Society through Barrow and Collins, and Newton seems to have discussed the possibility of an inverse square law of attraction for planets with Christopher Wren in 1677 when Wren was building the great library at Trinity College, Cambridge.[25] In 1673, Christiaan Huygens had published the v^2/r law for centrifugal acceleration and various people independently started coming up with inverse square laws after that time, though derived only for circular orbits.[26] Late in November 1679 Hooke, as Royal Society Secretary, wrote to Newton about the unsolved problem.[27] Hooke, perhaps partly influenced by Wren, had correctly defined the dynamics of orbital motion, assuming that Newton had not. There was, in fact, no centrifugal force at all; orbital motion was not a state of equilibrium between opposing forces; it was produced, instead, by the continued deflection of a body from a tangential path by a force acting towards a centre. In fact, Newton had already considered this way of looking at orbital dynamics, but Hooke's suggestion led him to look at the problem in an entirely new way, using Kepler's *second* law that equal areas are swept out by planets in equal times.

   Newton's reply was rather casual and offhand — he had only recently returned from Lincolnshire after his mother's death — and Hooke was delighted to spot what he thought was an error in Newton's remarks.[28]

---

[25] Newton to Halley, 27 May and 20 June 1686, *Corr.* II, 433–434, 435, 'I remember about 9 years since [around Easter 1677] Sr Christopher Wren upon a visit Dr. Done and I gave him at his lodgings, discoursed of this Problem of determining the heavenly motions upon philosophical principles. This was about a year or two before I received Mr Hooks letters. You are acquainted with Sr Christopher. Pray know when & whence he first learnt the decrease of the force in a duplicate ratio of the distance from the Center' (*Corr.* II, 434). The event recorded seems to have occurred just before Newton lost his most assiduous patron, the Master of Trinity, Isaac Barrow, of a 'malignant fever', on 4 May 1677.
[26] Huygens (1673/1929).
[27] Hooke to Newton, 24 November 1679, *Corr.* II, 297–298.
[28] Newton to Hooke, 28 November 1679, *Corr.* II, 301–302; Hooke to Newton, 9 December 1679, *Corr.* II, 304–306; Newton to Hooke, 13 December 1679, *Corr.* II, 307–308; Hooke to Newton, 6 January 1680, *Corr.* II, 308; Hooke to Newton, 17 January 1680, *Corr.* II, 313.

Again, Hooke's assumption was incorrect, for Newton's comment was valid for the case of resisted motion that he had assumed. Hooke, however, kept up the pressure, and wrote to Newton suggesting that an inverse square law would apply in the case of planetary motion, though he had no means of proving it and, though Newton did not reply, he had to satisfy himself that he could prove that an inverse square law would require elliptical orbits. Using this time, Kepler's second law, in a more generalised form, Newton was able to show that an elliptical orbit around an attracting body located at one focus required an inverse square law attraction. An autograph manuscript quoted by Brewster apparently had Newton claiming to have derived *Principia*, Book I, Propositions 1 and 11, which established the area law and the elliptical orbit, in December 1679.[29] The proof required calculus and so was entirely outside Hooke's range. Indeed, Hooke's own ideas, as Newton well knew, were hopelessly inadequate and confused, based as they were on his own fallacious law of forces and a fallacious law of velocities from Kepler.

Newton was at this time deeply into the study of alchemy, apparently convinced that he was on the verge of a major breakthrough. But he was sufficiently fired to investigate the comets of 1680–1682 and to attempt to work out the dynamics of their orbits. To obtain relevant data, he asked a Cambridge friend, James Crompton, to set up a correspondence with the Astronomer Royal, John Flamsteed, who had attended one of his lectures on mathematics in 1674.[30] The hundreds of MSS devoted to alchemical work, show that Newton probed the vast literature as never before or since. Despite the revulsion of Newton's Victorian biographer, David Brewster, for his subject's interest in the 'most contemptible alchemical poetry',[31] and despite the extravagant imagery and arcane terminology prominent in these writings — 'hunting of the green lion', 'doves of Diana', 'new-born king', and so forth — it seems that Newton was a 'rational' alchemist, determined to find the meaning behind the processes supposedly described in alchemical tracts,[32] and his own laboratory notes show a concern with quantitative analysis never before applied to this subject.[33]

---

[29] Brewster (1855, I, 272).
[30] Newton to Crompton, 28 February 1681, *Corr.* II, 340–347.
[31] Brewster (1855, II, 301).
[32] Newman (2002, 358–369).
[33] Westfall (1975).

The problem of planetary motion was apparently still unresolved, despite all the speculations, when, in January 1684, Christopher Wren, Robert Hooke and the new rising star of the astronomical world, Edmond Halley, met in a London tavern and made a wager in which Wren offered a valuable book to the first person who would demonstrate the relationship between an inverse square law and elliptical orbits. Hooke claimed success but failed to deliver the goods.

Then in the summer of 1684 Halley went to Cambridge to consult Newton himself:

> after they had been some time together, the D^r [Halley] asked him what he thought that Curve would be that would be described by the Planets supposing the force of attraction towards the Sun to be reciprocal to the square of the distance from it. Sr Isaac replied immediately that it would be an Ellipsis, the Doctor struck with joy & amazement asked him how he knew it, why saith he I have calculated it, whereupon Dr Halley asked him for his calculation without any farther delay.[34]

Newton was cautious and made out that he had lost the paper (conceivably, he hadn't) but he soon sent on a draft of a treatise on dynamics and its application to planetary motion which must have amazed even Halley.[35] Halley was a man of great personal charm and by some means or other persuaded the normally retiring Newton to launch himself into a full account of the new dynamics, based on a completely unprecedented universal law of force between all massive objects.

Throughout the composition Newton worked at a ferocious pitch, subjecting his psyche to the most extreme pressure and during this period he was in a volatile emotional state which required all Halley's skill to manage. Matters came to a head when Hooke laid claim to the inverse square law — how Halley managed to smooth this over we will never know. The result was the most extraordinary scientific book ever published, the *Principia Mathematica Naturalis Philosphiae*. The title was a deliberate inversion of Descartes' supposedly authoritative *Principia Philosophiae* of 1644.

---

[34]Testimony of A. De Moivre, Cohen (1971, Supplement 1.7, 297–298).

[35]Herivel (1965, 108–117) and Westfall (1980, 387–388), both supported the idea that the paper 'On motion in ellipses' (*Corr.* III, 71–76), later copied out for John Locke (*Unp.*, 293–301), was the original paper of 1679 or a version of it. Whiteside (*MP* VI, 553–554) and Hall (1992, 208–209, 236) remained opposed. In the autograph MS quoted by Brewster (1855, I, 272), Newton claimed that he added I, Propositions 6–10, 12–13 and 17, and II, 1–4, to the previously-discovered I, 1 and 11 in June and July 1684.

According to his amanuensis Humphrey Newton (no relation), Newton was during this period the very archetype of the distracted scholar:

'His carriage then was very meek, sedate & humble, never seemingly angry, of profound Thoughts, his Countenance mild, pleasant & Comely; I cannot say, I ever saw him laugh, but once.... He always kept Close to his Studyes, very rarely went a visiting, & had as few Visitors, excepting 2 or 3 Persons.... I never knew him take any Recreation or Pastime, either in Riding out to take the Air, Walking, bowling, or any other Exercise whatever, Thinking all Hours lost, that was not spent in his Studyes, to which he kept so close, that he seldom left his Chamber unless at Term Time, when he read in the Schools, as being Lucasianus Professor, where so few went to hear Him, & fewer that understood him, that of times he did in a manner, for want of Hearers, read to the Walls.' (A) 'When he read in the Schools, he usually staid about half an hour, when he had no Auditors he Commonly return'd in a 4th part of that time or less.' (B)

'I cannot say, I ever saw him drink, either wine Ale or Bear, excepting Meals, & then but very sparingly. He very rarely went to Dine in the Hall unless upon some Publick Dayes, & then if He has not been minded, would go very carelessly, with shoes down at Heels, stockins unty'd, surplice on, & his Head scarcely comb'd.' (A) 'Sir Isaac at that Time had no Pupills, nor any Chamber ffellow, for that I presume to think, would not in the least have been agreeable to his Studies. He was only once disorder'd with pains at the stomach, which confin'd Him for some days to his Bed, which he bare with a great deal of Patience & Magnanimity, seemingly indifferent either to live or dye. He seeing me much Concern'd at his Illness, bid me not trouble my self, for if, said he, I dye, I shall leave you an Estate, which he then mention'd.' (A)

'fforeigners He received with a great deal of ffreedom; Candour, & Respect. When invited to a Treat, which was very seldom us'd to return it very handsomely, freely, & with much satisfaction to Himself. So intent, so serious upon his Studies, that he eat very sparingly, nay, oftimes he has forget to eat at all, so that going into his Chamber, I have found his Mess untouch'd, of which when I have reminded him, would reply, Have I; & then making to the Table, would eat a bit or two standing, for I cannot say, I ever saw Him sit at Table by himself, At some {seldom} Entertainments the Masters of Colledges were chiefly his Guests. He very rarely went to Bed, till 2 or 3 of the clock, sometimes not till 5 or 6, lying about 4 or 5 hours, especially at spring & ffall of the Leaf, at which Times he us'd to imploy about 6 weeks in his Elaboratory, the ffire scarcely going out either Night or Day, he siting up one Night, as I did another till he had finished his Chymical Experiments, in the Performances of which he was the most accurate, strict, exact: What his Aim might be, I was not able to penetrate into but his Paine, his Diligence at those sett times, made me think, he aim'd at somthing beyond the Reach of humane Art & Industry.' (A)

The laboratory, 'On the left end of the Garden...near the East end of the Chappell', where 'at these sett Times, employ'd himself..., with a great deal of satisfaction & Delight', 'was well furnished with chymical Materials, as Bodyes, Receivers, ffends, Crucibles &c, which was made very little use of, the Crucibles excepted, in which he {fused} his Metals: He would sometimes, thô very seldom, look into an old mouldy Book, which lay in his Elaboratory, I think it was titled, — Agricola de Metallis, The transmuting of Metals, being his Chief Design, for which Purpose Antimony was a great Ingredient.' (B) 'Nothing extraordinary, as I can Remember, happen'd in making his Experiments, which if there did, He was of so sedate & even Temper, that I could not in the least discern it.' (A)

'He was very Curious in his Garden, which was never out of Order, in which he would, at some seldom Times, take a short Walk or two, not enduring to see a weed in it...' (A) The garden 'was kept in Order by a Gardiner I scarcely ever saw him do any thing (as pruning &c) at it himself. When he has sometimes taken a Turn or two, has made a sudden stand, turn'd himself about, run up the stairs, like another Archimedes, with an *Εὕρηκα*, fall to write on his Desk standing, without giving himself the Leasure to draw a Chair to sit down in. At some seldom Times when he design'd to dine in the Hall, would turn to the left hand, & go out into the street, where making a stop, when he found his Mistake, would hastily turn back, & then sometimes instead of going into the Hall, would return to his Chamber again.' (B) 'He very seldom went to the Chappel, that being the Time he chiefly took his Repose; And as for the Afternoons, his earnest & indefatigable Studyes retain'd Him, so that He scarcely knewe the Hour of Prayer. Very frequently on Sundays he went to Saint Mary's Church, especially in the fore Noons....' (A)[36]

Halley organised the publication of the work with the Royal Society to give it a semi-official status, obtaining the imprimatur of the then President, Samuel Pepys, in June 1686, and dedicating the book to King James. Unfortunately, the Royal Society had bankrupted itself by publishing Francis Willughby's *History of Fishes*. Although Halley had recently resigned his Society fellowship to obtain a paid position as a clerk, he undertook the publication costs himself, and was paid his salary in unsold copies of the *History of Fishes*. It is impossible to overestimate the debt that the world owes to this action; Halley must be given full credit for realising that Newton's 'divine treatise' was the opportunity of a lifetime.[37] No great scientific discovery has ever been established without the selfless support of such able lieutenants.

---

[36](A) Humphrey Newton to Conduitt, 17 January 1728, KCC, Keynes MS 135, NP; (B) 17 February 1728, KCC, Keynes MS 135, NP.
[37]Halley to Newton, 5 April 1687, *Corr.* II, 473.

The *Principia* was immediately seen as a stupendous achievement. Huygens, whose opinion mattered more than most, was greatly impressed, though he demanded a 'physical' explanation of gravity,[38] and Leibniz used it to restructure his own approach to mechanics. Though a Scholium to Book I, Proposition 4 gave quite generous credit to Wren, Hooke and Halley for their roles in the gestation of the law of gravitation, Hooke, true to form, believed that he hadn't received sufficient acknowledgement, and refused to speak to Newton again. He gained few supporters, for his record as a universal 'claimant' told against him, not to mention his well-known inability at formulating a coherent dynamics.

Yet, for all its subsequent celebrity and success, part of the significance of the book was missed entirely. It is possible to argue that Newton's successors confused the shadow and the substance, and latched on to the synthetic mathematical wizardry of the *Principia* while being unaware of its ruthlessly abstract and analytical *qualitative* basis. Newton's mathematical approaches were used for mechanistic schemes which he would never have entertained. His successors admired the text while missing the significance of the proto-modern (and, at the same time, neo-mediaeval) subtext.

While the *Principia* was in process of publication, Newton was putting his entire career at risk in a dispute with the monarchy over the granting of a Cambridge degree to a Catholic priest, Alban Francis; this included a memorable encounter with the notorious Judge Jeffreys, and stood him in good stead after the political revolution of 1689. On 15 January 1689, Newton was elected Member of Parliament for the University. This was the 'Convention Parliament' that completed the 'Glorious Revolution' by announcing on 22 March that King James II had abdicated, and offering the throne to William and Mary. Newton spent more than a year in the capital until Parliament was dissolved on 6 February 1690, and began to realise that he had unexpected possibilities as a member of London Society. For a few years after 1689 he had hopes of a young disciple, Nicholas Fatio de Duillier, who developed new theories of the aether, but these hopes remained unfulfilled. Despite being a celebrity, positions in London were not immediately forthcoming, and Newton in the meantime made the most prodigious effort to extend his work in every conceivable direction, so much so that, in October 1693, he had something of a nervous breakdown, writing

---

[38] A very strict Cartesian, Huygens was only convinced of the correctness of the elliptical orbits of planets and Kepler's area law after he had read Newton's book. Newton's realisation of the significance of the first two Keplerian laws was an important achievement in itself.

letters to his friends Locke and Pepys accusing them of plotting against him.[39] The reasons for this have never been properly established, but a possible contributory factor may have been a catastrophic alchemical failure. Some of the ideas of these years are among his most remarkable achievements and were incorporated later into his optical Queries.

After several years of lobbying he became Warden (1696), then Master (1699), of the Royal Mint, then based in the Tower of London, and proved himself to be an outstanding administrator. He played a major part in the great recoinage of 1697, which may have saved England's economy from bankruptcy. During his period as Warden, and particularly during the critical months of the recoinage, he embarked on a ruthless campaign against several leading members of the London underworld, giving absolutely no quarter to people who were regarded by the authorities (with a good deal of justification) as dangerous and unprincipled criminals. There is no reason to suppose that he regarded this as anything other than a duty of the office, one, in fact, that he tried to avoid.[40] It is extremely unlikely, as Frank E. Manuel tries to imply in his provocative but highly speculative *Portrait of Isaac Newton*, that he was using the opportunity to vent his considerable spleen on what might be taken safely as a righteous cause.[41] In 1703, when Hooke was safely out of the way, he accepted election as President of the Royal Society. It has been claimed that Newton's term as President was the cause of the disappearance of Hooke's portrait from the Royal Society's collection, but this is only speculation. It is certainly not an established 'fact', and it is far more likely that the portrait never existed.[42] The *Scheme for establishing the Royal Society*, which he drew up as incoming President in 1703, without in any way mentioning Hooke by name, shows that he at least valued what Hooke had done in his office as curator of experiments.

He retained the office of President and his position at the Mint until his death in 1727; he was MP again for Cambridge University in 1701–1702, and was knighted, for political reasons, in 1705. During this period it has been claimed that he allowed his political patron, Lord Halifax, to have an illicit affair with his niece, Catherine Barton, who lived with him after he moved to London, an issue that was guaranteed to shock the Victorians, if it hardly bothers us. Yet, as with many such gossipy rumours, there isn't the slightest

---

[39] Newton to Locke, 16 September 1693, *Corr.* III, 280; Newton to Pepys, 13 September 1693, *Corr.* III, 279.
[40] Newton to the Treasury, July/August 1696, *Corr.* IV, 209–210.
[41] Manuel (1968, 229–244).
[42] Chapman (2004). Hooke had died on 3 March 1703.

direct evidence that it is true, and very possibly it isn't.[43] As President, Newton did much to revive the fortunes of the Royal Society which had gone into a sharp decline during Hooke's long illness in the 1690s, and he revived the tradition of performing experiments at the meetings, hiring the outstanding experimentalist Francis Hauksbee as the first demonstrator.

Far from giving up research, as is often supposed, Newton pursued entirely new questions concerning electric, magnetic and capillary forces and developed many startlingly original and penetrating ideas. In 1704 there appeared, at long last, the *Opticks*, the extended account of his optical researches, and the first of his mathematical tracts. The *Opticks* was concluded by a series of 16 Queries, which was extended to 31 by the time of the third edition in 1717. These Queries reveal a great deal more about Newton's characteristic thought patterns than can be seen in the *Principia*. But they are merely a small fragment of a vast amount of manuscript material which shows Newton's special capacity as a qualitative thinker. Even in their truncated and published form, they were extremely influential, especially in 18th century chemistry.

Despite Newton's success with the *Principia* and *Opticks*, the 'revolution' was not yet accomplished. Cartesianism was still the dominant cosmical philosophy on the Continent of Europe in the early 18th century, and was even being taught at Cambridge. Newton's optical discoveries were still a matter of doubt until the then Royal Society demonstrator Desaguliers demonstrated them to a party of French visitors in 1715. Great battles had still to be fought, and not always scientific ones. Newton's natural philosophy had to be established, not just on its merits as a system, but on the overwhelming impact of its total fire power. After the experience of 1672, Newton no longer expected the obvious to be recognised immediately by the best scientific minds. Other methods had to be employed in a concerted campaign. A dispute with Flamsteed, the Astronomer Royal, over the publication of his astronomical observations, led to the creation by Newton of a Board of Visitors to Greenwich Observatory, and the compulsory publication of Flamsteed's star catalogue in 1712.

Though Newton had now gained fame as the author of the *Principia*, the experience over the optical paper had a profound effect on him. In his later work he suppressed much, smoothed over difficulties and gave the impression

---

[43]Most biographers leave the verdict open, but Hall (1992, 302–305), expresses considerable doubt. David Berlinski's distinctly post-Victorian view is that: 'It was a drama that did no one harm and almost everyone credit.' (2001, 167) Voltaire (1757) is the source of the early rumours. Other early accounts can be found in Brewster (1855) and de Morgan (1885).

that the work was much more authoritative than it actually was. With his basic philosophy of gravity under attack from the Continental Cartesians, he was not in a position to do anything else. Having made the claim that his system was universal, he could not afford to be wrong on any issue, even one of detail, and felt obliged to overstate the accuracy of his work, especially in the test of the moon's gravity, the precession of the equinoxes and the calculation of the velocity of sound. It has been claimed that, in these instances, Newton 'fudged' the data to make them look more accurate fits to his theory.[44] However, this no longer seems to be a reasonable assessment in at least two of the cases, if not all three, and he certainly didn't fabricate evidence.

In his old age, he worked hard on his prophecy and chronology. During this period, he extended his interests in prophecy to the study of Daniel as well as Revelation, which he believed to be divinely inspired in a way that the historical books of the Old Testament were not. All the same, Newton had inevitably become much more worldly-wise during these years; and in one of his manuscripts on the Apocalypse of St John, he made the remarkable slip of writing Sr John for St John![45] 'The man who had once prepared to surrender his fellowship not to accept the mark of the beast now cultivated the odor of orthodox sanctity by serving as a trustee of the Tabernacle in Golden Square and as a member of the Committee to Build Fifty New Churches in London.'[46] He may have been seldom in chapel in Cambridge, but in London he made sure he was seen going to Church every sabbath.

There was a rather nasty moment when Newton's successor at Cambridge and protégé, William Whiston chose to make a public declaration of his unitarian beliefs in October 1710. Whiston had naïvely expected Newton to come to his support, but the latter knew that any whisper of a threat to the established religion would lead to immediate condemnation of both himself and his whole world system. Science may well have suffered the same fate in England as it did in Italy after Galileo.

Newton was well advised in his silence, even if this meant abandoning his protégé to the wolves, who had him ceremoniously drummed out of Cambridge. England was then in the throes of what was probably the last of its public rituals of institutionalised religiosity. Earlier in the year, Dr. Henry Sacheverell, in search of preferment, had made a covert attack

---

[44] Westfall (1973). Cohen (1999, 361–362) has already refuted the charge in relation to the velocity of sound.
[45] Westfall (1980, 814).
[46] Westfall (1980, 814–815).

on the predominantly Whig government by preaching a sermon attacking religious toleration, and accusing the new academies, which had been set up to educate religious dissenters, for teaching 'all the hellish principles of Fanaticism, Regicide, and Anarchy.' The government blundered by impeaching Sacheverell for seditious libel and was brought down on a wave of popular support for this High Church rabble-rouser. It was not a good time to be a non-conformist of any kind, and, least of all, anything that might be considered the next best thing to a deist or atheist. Newton's brilliant lieutenant, Samuel Clarke, who was soon to defend his master against the insinuations of Leibniz that the Newtonian system was conducive to atheism, was implicated with Whiston but saved his position by perjury. Newton forever had to walk a tightrope of fear that his true beliefs would be found out, and at the same time reconcile his conscience with the public conformism which he had no choice but to adopt. He made his own gesture by refusing the sacrament on his deathbed in 1727, but this was not widely known at the time.

This was the background to the bitter quarrel with Leibniz, which was ostensibly concerned with the invention of the calculus, but which was really a clash between two incompatible philosophies. The dispute began in earnest with an accusation in a book by Fatio, published in 1699, that Leibniz had stolen his version of the calculus from Newton,[47] but Newton avoided any direct involvement for a number of years after this. The long-running dispute, mostly conducted by parties close to the two men, was ended by a Royal Society report in favour of Newton, which Newton effectively drafted himself, and published as the work of the Society in February 1713.[48] For good measure, he also published an ostensibly anonymous review of the report, again written by himself, in the *Philosophical Transactions*, early in 1715.[49] At the same time, he was working on a second edition of the *Principia*, which was published in 1713.

After the third edition of the *Opticks* in 1717, Newton did very little new work in physics, except for a third edition of the *Principia* which was published in 1726. He worked mostly at his *Chronology of Ancient Kingdoms Amended* and *Observations on the Prophecy of Daniel and the Apocalypse of St John*, which were published after his death.[50] He became very wealthy, and remained so, though he lost as much as £20,000 in the South Sea Bubble

---

[47] Fatio (1699).
[48] *Commercium Epistolicum*, 1712, issued 1713.
[49] 'Account', *Philosophical Transactions*, 1715, 173–225, written 1714.
[50] In 1728 and 1733.

of 1720. In his last years, he suffered from the stone and other ailments, and retired from central London to Kensington, which was still semi-rural. He died on 20 March 1727 and was buried in Westminster Abbey under a magnificent monument.

2.3. Newton's Character

One of the biographic clichés of modern times is that great personal achievement can only occur at the same time as a corresponding personal cost. In an age of democratic aspiration, we can only tolerate heroes if we can see that they have compensating flaws which make them 'human' like the rest of us. Of course, no one is above criticism but something more is involved in the process. Biographers of Newton have described his character in terms which reflect the interests of their own day. The 19th century required a patron saint, and so we had the saintly Newton, a paragon of virtue above petty considerations, dedicated to the progress of rational scientific truth. The 20th century liked its heroes flawed and so we had the neurotic Newton, full of angst and hovering permanently on the verge of a nervous breakdown — a man whose rational scientific work was a mere side-issue within a vast and complex substructure of ideas on theology, prophecy, chronology and alchemy, which make very strange reading today. Added to this came a gradual emphasis on the darker side of Newton's character, the secret unitarian with a tormented Puritan conscience, the ruthless political manipulator, the merciless persecutor of clippers and coiners — images exploited to the full by those who stress the ruthless, impersonal, hard-edged or 'monolithic' character of modern science and its abandonment of human virtues and ethical standards.

One of the main sources of this picture, to which we have already referred, was Manuel's very eloquent and fascinating *A Portrait of Isaac Newton*, published in 1968. Manuel was a very considerable scholar, who opened up the subject of Newton as a theologian and chronologist to serious scrutiny for virtually the first time. However, he was also afflicted with one of the major curses of 20th-century biography, the belief that Freudian analysis can give credible insights into a subject's behaviour and personality. Manuel paints a picture of a man with a Messianic complex, a man who dared to anagrammatise his name as IEOUA SANCTUS UNUS, and one who rejected the Trinity to (subconsciously) replace Christ as God's true prophet with himself.[51] It is a marvellous fantasy which one is tempted to believe

---

[51] In alchemical notebook, Westfall (1980, 289); Gjertsen (1986, 17).

contains some element of truth, but, ultimately, it is too much of a projection on Newton's character of the author's views on modern science. Manuel's account had a serious effect on nearly all subsequent biographers, however much they tried to avoid his influence, and it is hard to deny that Manuel's prose is mesmerising and hypnotic. However, it is another thing to claim that it is the truth. It appeared when even real scholars still widely believed that Freudianism had scientific validity, before more credible approaches exposed the ultimately pseudoscientific nature of the whole enterprise.

Manuel's account still influences the popular view of Newton's character, but the elements of fantasy make it, in these respects, no more credible and amenable to rational argument than the Newtonian studies of prophecy and chronology which he so ably explicated. Newton was undoubtedly a most remarkable man, possibly possessed of the most powerful intellect in the whole of human history, and from a remarkable man we may well expect some remarkable behaviour, whether good or bad. But the exaggerated pictures that followed on from Manuel have not helped to establish what kind of man he really was, and psychological stereotypes have almost zero explanatory value when we are confronted with such a complex personality.

No man of Newton's intellect is likely to represent a median on the scale of human personality, but he was certainly within the spectrum of normality, if perhaps nearer one end of it than the centre. He didn't come over to people who met him as a 'genius', in the popular sense of the term. He was 'of no very promising aspect', according to Thomas Hearne, and he 'did not raise any expectations in those who did not know him', according to Bishop Atterbury.[52] He had no spontaneous verbal skills and, as a public servant, tried to avoid situations where he had to give testimony without using a prepared statement. During two terms in Parliament, he never once contributed to the debates. A famous story has the prominent French mathematician, the Marquis de l'Hôpital, asking Dr. John Arbuthnot: '...does he eat & drink & sleep. is he like other men?' To which Arbuthnot replied that 'he conversed chearfully with his friends assumed nothing & put himself upon a level with all mankind.'[53] Clearly, most people didn't think he behaved in a way that suggested abnormality. Godfrey Kneller painted a very striking portrait in 1689, but by then everyone knew Newton as the author of the *Principia*, and Kneller, who painted hundreds of purely routine portraits during his career, was generally stirred to greatness by subjects he knew were of special interest.

---

[52] Both in Brewster (1831, 342).
[53] Keynes MS 130.5, Sheets 2–3, Westfall (1980, 473).

Newton was obviously a driven man. He seems to have realised early on that he had a mission and was prepared to sacrifice everything else, including his health, physical and mental, to complete it. Apart from an unquenchable appetite for work, his was not an addictive personality. His appetites, such as they were, were kept firmly under control. He was sparing in his consumption of food and drink, and, for health reasons, became a vegetarian in later life. He avoided things such as tobacco because, he said, 'he would make no necessities to himself',[54] but he had a range of quack remedies for various illnesses, which he was always willing to share with anyone who was willing to listen. He was secretive (though often with good reason), neurotic but not psychotic, and neither autistic nor bipolar, but a celibate with no known sexual encounter or interests (and there is no reason for us to make them up[55]).

He was an incredible polymath and a prodigious worker much beyond the capacity of ordinary men but perhaps more typical in those of equally restless intellect. He was immensely versatile, his earliest investigations involved a phonetic alphabet and a universal language.[56] For him, the study of all human experience was one great enterprise. Determined to see the action of God's providence in the universe, he studied history and prophecy at least as intently as he studied natural philosophy. He brought the same methodological principles to all his studies, and was always rational in his approach, though in some areas this produced highly non-rational results through his unfounded belief in the credibility of dubious sources. Newton lived in an age when men had strong religious opinions and his whole manner of life was determined by intense religious beliefs, in a way quite alien to the

---

[54] Gjertsen (1986, 178).

[55] The only direct testimony to anything that might resemble love interest is the word of Catherine Storer (later Vincent), who in later life recalled that Newton had had a fondness for her in youth. He later visited her in the country and presented her with a sum of money when she needed it. Stukeley (1752, 46). The identification is in Christianson (1984, 583). Newton, of course, would not have been eligible to marry during the long period of his fellowship at Trinity College, Cambridge, from 1667 to 1701. The idea of a 'relationship' between Newton and Fatio beyond their short-lived connection as master and pupil is another one that had its genesis in Manuel's work, and it has been duly followed up in some of the more popular biographies, such as White (1997). However, the evidence as presented doesn't add up to a case beyond the speculation characteristic of a post-Freudian and post-Kinseyan period. Gjertsen (1986, 205) says that even Newton's enemies found nothing 'worthy of comment' in the interactions he had with Fatio and other pupils.

[56] *Of an Universal Language*, c 1661, Elliott (1957); *A Scheme for Reformed Spelling*, 1661–1662, Elliott (1954).

ethics of the 21st century. In this, he reflected his time and was only different from other men of his century in his unwillingness to compromise.

No one would say that Newton was the most attractive character who ever lived. He was not, like Michael Faraday, of a saintly disposition, and he was no genial companion, like James Clerk Maxwell or Edmond Halley, 'but the few admitted to friendship found a lively and hospitable companion.'[57] Humphrey Newton said that: 'When he was about 30 years of Age, his gray Hairs was very Comely, & his smiling Countenance made him so much the more graceful.'[58] He certainly had friends and seems to have kept on the right side of a number of men of good or otherwise agreeable character, such as Henry More, Isaac Barrow, Boyle, Wren, Locke, Huygens and Halley, as well as Stukeley and Conduitt in later life. He appears to have had a genuine friendship with Samuel Pepys, who, as a character, was almost his polar opposite. He was also prepared to take due responsibility for providing for members of his family, while the presence of his vivacious niece, Catherine Barton, livened up his London household, perhaps for as long as 20 years. There was, nevertheless, a notable lack of love in his life and seemingly precious little humour, though he is described in some accounts as being 'very merry' on occasions later in life.[59] His correspondence reveals an almost unrelieved seriousness, though one letter, which Westfall believed to be a forgery, shows an unexpected vein of light irony.[60]

According to William Whiston, who was prejudiced, he had 'the most fearful, cautious, and suspicous Temper, that I ever knew',[61] while John Locke, who was not and who regarded Newton highly, thought that he was a 'nice' (that is, awkward) man to deal with.[62] Humphrey Newton, however, thought that 'His Behaviour was mild & meek, without Anger, Peevishness or Passion, so free from that, That you might take him for a Stoick.' And that he was 'No way litigious, not given to Law or vexatious suits, taking Patience to be the best Law, & a good Conscience the best Divinity.'[63] Newton was never a man who sought out disputes (unlike some of his later opponents); the unpleasant side of his character only came out when he was forced into them.

---

[57] Hall (1992, xiv).
[58] Humphrey Newton to Conduitt, 17 February 1728, KCC, Keynes MS 135, NP.
[59] Humphrey Newton said he was 'very merry' when an acquaintance asked him what use studying a copy of Euclid would be. The story is recorded in Stukeley (1752, 57).
[60] Newton to Maddock, 7 February 1679, *Corr.* II, 287–288; Westfall (1980, 279).
[61] Whiston (1753, Vol. 1, 250).
[62] King (1830), Vol. II, Locke to King, 38.
[63] Humphrey Newton to Conduitt, 17 February 1728, KCC, Keynes MS 135, NP.

While there is no evidence of great personal warmth, he could be 'steady, resolute and generous'; he indulged in a number of charitable acts and made considerable financial gifts to needy members of his family and others. He 'gave away large sums and pushed the careers of young men of whose ability he was convinced.'[64] For Humphrey Newton:

> He was very Charitable, few went empty handed from him. Mr. Pilkinton, who liv'd at Market-Orton, died in a mean Condition (thô formerly he had a plentiful Estate) whose Widow with 5 or 6 Children Sir Isaac maintain'd several years together. He commonly gave his poor Relations (for no ffamilies so rich, but there is some Poor among them) when they apply'd themselves to him, no less then 5. Guineas as they themselves have told me. He has given the Porter many a Shilling, not for leting him at the Gates at unseasonable Hours, for that he abhor'd, never knowing him out of his Chamber at such Times.[65]

There are many more polite and reasonable letters in his correspondence than there are violent bursts of temper. He was generally patient in answering genuine queries on scientific matters, and, as the story by Arbuthnot suggests, notably lacking in arrogance. Humphrey Newton, as we have seen, described his manner during the composition of the *Principia* as 'meek, sedate and humble.'[66]

For all his many years of isolation at Cambridge as a scholastic recluse, Newton still managed to pass off as a more or less acceptable member of London society for over 30 years. The treacherous Antonio Conti, coming armed with prejudice from the Leibniz camp, found that he rather liked him.[67] People behave differently in different circumstances and at different times in their lives; a person's character is not necessarily fixed throughout life — otherwise there would be no divorce! The recluse writing the *Principia*, described in Humphrey Newton's remarkable account, was obviously very different from the public figure of the London years, and, like many who reach an advanced age, he seems to have become a rather more genial figure in his final years, once he had outlived all his enemies.[68]

As an enemy he had a slow-burning, rather than a short, fuse, but was capable of occasional volcanic eruptions. He reacted particularly strongly when he felt his character was being impugned, as Anthony Lucas, William

---

[64] Hall (1992, xiv).

[65] Humphrey Newton to Conduitt, 17 February 1728, KCC, Keynes MS 135, NP.

[66] Humphrey Newton to Conduitt, 17 February 1728, KCC, Keynes MS 135, NP.

[67] Christianson (1984, 544).

[68] In the words of Buchwald and Feingold, 'the introverted young man grew into the socially adept and ever-vigilant president of the Royal Society' (4).

Chaloner and Leibniz, in particular, would find out. Highly strung, he suffered one nervous breakdown but quickly recovered. Newton could be stern and harsh in dispute but he was seldom the first aggressor; he was a ruthless adversary, and never gave any quarter, but he usually sought to avoid confrontation, and was content for long periods to let sleeping dogs lie. Much has been made of the fact that he had three major quarrels in his life, but Hooke and Flamsteed, two of his main opponents, had more, and both could be disagreeable or cantankerous. Neither, however, has received as much criticism.

According to one of his biographers, Rupert Hall, who wrote after Manuel and Westfall, the age of Newton was one in which 'the defence of intellectual property was essential to success and strong writing on either side of a point in dispute usual.' In such a climate, 'Newton's ill manners rarely seem to exceed the norm.'[69] He found himself thrust into this world on his first publication in 1672, and it carried on without any let up until his last, the pirated *Abrégé de la chronologie*, in 1725. Every one of his publications during this period was attacked and vilified on one ground or another. His general attitude towards his correspondents seems to reflect their attitudes to him. What Westfall considered to be 'paranoia' in Newton's 'brutal' last replies to the Jesuit Anthony Lucas (which may never have been actually sent and were certainly never published) seems to have been no more than justified exasperation with a correspondent who was both accusatory and discourteous and who continually shifted the goalposts in his arguments. Lucas, in particular, made the cardinal error of accusing Newton of producing false reports of his work.[70] Newton made his feelings clear when John Aubrey wrote that another letter awaited him in London, 'Pray forbear to send me anything more of that nature.' He reacted in exactly the same way as anyone would with a persistent and uninvited salesman.[71] Hall says that 'A tendency to classify Newton as clinically ill might perhaps be diminished by greater familiarity with other examples of coarse and 'brutal' language in 17th-century controversies.'[72] And people who imagine Newton as being

---

[69] Hall (1992, 140). Gorlin (2008–2009) has written a very effective refutation of Manuel.

[70] Newton to Lucas, 5 March 1678, *Corr.* II, 254–260, 262–263; Westfall (1980, 278–279); Hall (1992, 140). The correspondence with Lucas seems to have been one of a number of incidents in which Westfall's interpretation of Newton's character was unduly influenced by Manuel's Freudian interpretation.

[71] Newton to Aubrey, c June 1678, *Corr.* II, 269. Westfall has proposed that Newton had a kind of mini-breakdown in late 1677 as a result of a fire in which he lost a work on optics, some fragments of which still survive.

[72] Hall (1992, 140).

singular or in a minority in adopting such a style have obviously not read such things as the polemical works of the sublime poet, John Milton.

In connection with priority battles, Newton seems to have been much less concerned with his 'intellectual property' rights than someone like Hooke, and it is not at all true that he didn't attribute work to his predecessors. There are many citations in his works of earlier and contemporary scientists — even to Hooke and Flamsteed, though these were reduced in later editions, as the quarrels developed.[73] Neither felt that they had received enough credit, but it is difficult to believe that they would have been easily satisfied by even more handsome acknowledgements. Newton sometimes attributed results to other authors, even when there was no need. He probably hadn't read Copernicus or Kepler, so he would have been unclear as to their contributions. As it was, he was perhaps a little too generous to Galileo (thereby creating confusions in the historical record) and perhaps not generous enough to Kepler in not specifically naming him in the *Principia* in connection with the first two Keplerian 'laws' that he used, though he did name him elsewhere in connection with the 'first law', of elliptical orbits, and the 'second law' of equal areas,[74] and he named Kepler as the author of the 'third law' in Phenomenon 4 and the Scholium to Proposition 4 of *Principia*, Book III. And, of course, he even drafted passages giving away the entire theory of gravity to the ancients!

Both Johann Bernoulli and, to a perhaps lesser extent, Gottfried Wilhelm Leibniz, with whom he quarrelled over the invention of the calculus, were guilty of seriously duplicitous conduct on more than one occasion. Both (unlike Newton) have been shown as resorting to plagiary at different times in their careers, and their attacks, if left uncountered, would have been damaging to Newton's reputation. In fact, much of the negative analysis of Newton's character stems ultimately from the revelations which began to emerge in the 19th century showing that he had manipulated his position at the Royal Society to brand the apparently 'blameless' Leibniz as a plagiarist. It is dangerous, however, to take up moral stances on the basis of evidence which is only partial when more extensive investigations might lead to a totally different picture. We now know that Leibniz was far from blameless,

---

[73] He was particularly strong in his praise of Huygens, in whom he recognised a great geometer like himself.

[74] First law, letter to Halley, 20 June 1686, *Corr.* II, 436, where he says that Kepler 'guest' the orbits 'to be Elliptical', draft letter to Des Maizeaux, c 1718, Cohen (1971, 295); second law, *De Motu Corporum in Gyrum*, Problem 3, Scholium, MP VI, 48–49, Phaenom. 14, *Phaenomena* (MS), *Unp.*, 385. Smith (1996, 357) points out that all three 'laws' were 'in some dispute' in the 1680s.

and, despite all the detailed studies which have been made of the priority dispute, we still don't have unequivocal evidence of what really happened to trigger it. The last word has yet to be written on this subject.[75]

Certainly, most of the actions Newton took have logical explanations. None were simply a reflection of his personality or emotional state. They were always a response to circumstances, which demand to be understood within their original context and not according to some supposed absolute standards of morality or good conduct formed under quite different conditions in a totally different era. In particular, Newton decided that duplicity by his enemies would be answered by even more subtle duplicity by himself, and it is difficult to imagine how anyone would have been able to effect such a vast realignment of human thinking as he ultimately accomplished without at least a streak of the steely ruthlessness he showed in the controversies which such a shift inevitably generated. Critics who have been quick to pass judgement have been slow to tell us what alternative courses they would have taken in the same situation, for they are far from obvious. Sometimes it is not possible to do a 'good' action in particular circumstances. Many authors have painted a picture of Newton as authoritarian, but his correspondence doesn't give this impression. Also, people don't try it on with those who adopt a strongly autocratic style. They tend to do this with those who, like Newton, attempt at first to be more subtle.

Confronted with any foe, whether public or private, Newton fought to win when he realised that he had to come to blows with his adversaries, and win, he always did. Conflict is always unpleasant and diminishing even in victory, but it can be more damaging trying to avoid it. Newton was certainly merciless as a public servant to those who were declared enemies of the state, but only to the extent required of his office. He was utterly reliable, and willing to carry his duty to the limits; he didn't actively seek unpleasant duties, but didn't shirk them either. Manuel made much of the way he dealt with clippers and coiners as Warden of the Mint, even to hounding them to their trial and execution. There is not the slightest evidence that this filled any personal need in Newton. The coiners of the 1690s were the equivalent of drug dealers today, prepared, in the case of a man like William Chaloner, to encompass the deaths of others in pursuit of their crime. The much more genial Henry Fielding had exactly the same attitude to criminals half a century later, as did Pierre de Fermat at an earlier period. Many people regard the death penalty as harsh or uncivilised, even for the most horrific

---

[75]The issue will be discussed in *Newton — Innovation and Controversy*, the final volume in this series.

of crimes; but every society that has ever existed has been prepared to make decisions on life and death to protect itself.[76]

To a large extent, Newton's activities in the apprehension and prosecution of coiners fulfilled a role within a larger process, which always involved a considerable number of other individuals. The process had begun before his time at the Mint and carried on long after him. However, Newton took the pursuit and eventual conviction of Chaloner, a 'master criminal' and effectively a double murderer, as a personal responsibility, to which he devoted many hours of effort over a period of more than 2 years. Once Chaloner had accused Newton himself of counterfeiting, in testimony given in February 1697, and tried to usurp his position at the Mint, his fate was sealed. He was pursued with every means possible, using informers, spies and witnesses, true or false, hoping for their own pardons. An anonymous twelve-page pamphlet on the Chaloner case, entitled *Guzman Redivivus* (1699), has been described as among 'the more curious and colourful publications of the late 17th century', and has even been attributed by one of Newton's biographers, Gale E. Christianson, to Newton himself.[77] If so, it would be very different in character from any of his other writings. Newton, however, and everyone who had had any dealings with Chaloner would, have agreed with its summary of his position in society 'as a rotten Member cut off.' Not a single individual who had any knowledge of his many criminal activities had the slightest interest in his pleas for justice and merciful treatment.[78]

Conforming with a stereotype associated with a particular kind of scientist, Newton does not seem to have shown any great aesthetic sense, though some is evident in his response to colours in the *Opticks* and optical manuscripts,[79] and he had a childhood interest in painting and drawing. At the end of his life, he owned a considerable collection of prints and paintings. He was the subject of several striking portraits, and designed a number of medals for the Mint, as well as producing a reconstruction

---

[76] Levenson considers Newton's attitude to crime and punishment to be very much the norm for his age, a long way from Manuel's fevered speculations. He describes Manuel's speculations and those of authors who follow this line as 'nonsense' (2009, 165), and says that 'the record of [Newton's] depositions shows him to be simply a relentless practical man doing his job' (http://www.executedtoday.com/2009/03/22/1699-william-chaloner-isaac-newton-counterfeiter/).

[77] Christianson (1984, 405–406).

[78] Levenson (2009) provides an entertaining account of Newton's career as a detective.

[79] He refers in a draft Q 16 to 'the colours of bodies where in the beauty of nature chiefly consists' (CUL, Add. 3970.3, f. 234^v, NP) and asks in another draft Query: 'Whence is it that Nature doth nothing in vain & whence arises all that beauty that we see in the world?' (*ibid.*, f. 247^r, NP).

of Solomon's Temple as a monument of sacred architecture. Though he discussed harmony in mathematical and scientific terms, he is not known to have been interested in music. He is said to have visited the opera once, and enjoyed the first act, but was glad to escape before the end.[80] Opera, however, is an extravagant form, and the story cannot be taken as proof that he disliked music. He wrote fascinating prose, sometimes touched with the poetic, as in *An Hypothesis of Light*, but had no obvious interest in literature. He had just two plays by Shakespeare in his library, *Hamlet* and *The Tempest*, though these are among the most significant, and only the 1720 edition of Milton's *Works*. His considerable collection of classical literature seems to have been mostly a source of information on ancient knowledge. There is, however, an aesthetic sense in science, which he seems to have felt, and Chandrasekhar and other commentators often describe his mathematical results as 'beautiful'.[81] And while the *Principia* is usually seen as the ultimate science classic, the *Opticks* is often regarded as a book with an equally aesthetic appeal. Einstein wrote of its author as an 'artist', while John Heilbron thought he could even be considered 'romantic'.[82] Ultimately, Newton emerges as a man who certainly had flaws, though not necessarily to a particularly exceptional degree, and certainly not to the degree assumed by sensationalist biographers. His singular importance in the history of science shouldn't lead us to exaggerate them to compensate for having to acknowledge his towering intellect.

---

[80] Stukeley (1752, 14).

[81] For example, Chandrasekhar (1995, 201). Chandrasekhar manifests a distinctly aesthetic response to many other theorems in the work.

[82] Einstein (1931/1952, lix): 'In one person he combined the experimenter, the theorist, the mechanic, and, not least, the artist in exposition.' Heilbron (1982, 43): 'The tight, self-justifying, towering mathematician of the *Principia* seldom appears in the more open, accessible and even romantic author of the *Opticks*.'

Chapter 3

Waves

3.1. The Theory of Wave Motion

The study of wave motion lies at the foundation of most of physics, and is as intrinsic to quantum mechanics as it is to classical physics. Newton is not usually strongly associated with the analysis of wave motions, yet he was effectively its creator, initiating the modern mathematical theory of waves, and of simple harmonic motion, and showing their connection. The ideas he used are exactly those we use for wave theory today, after they were put into their modern algebraic form by Euler by direct transposition from the *Principia*.[1] Though Newton's mathematical analysis of waves had an indirect connection with his work on optics (and ultimately led to the 19th century breakthrough in the mathematical theory of that subject), it was included in the *Principia* rather than the *Opticks*, though interestingly only in the more risky and speculative second book. It was a notable example of a dynamical analysis with no direct connection to gravitation or the theory of planetary motion.

For Newton, as for physicists now, wave motion could be seen as generating or being generated by simple harmonic motion of the component particles of a medium. Simple harmonic motion was already of interest during the period as generating the isochronicity (or period regularity) of such time-measuring devices as the simple pendulum and the balance spring, and for its connection with the cycloid curve. Newton gave the first general mathematical analysis of this subject in his discussion of fluid motion in the second book of the *Principia*.

---

[1] Westfall (1971, 497) considers it a 'supreme irony' that 'the champion of the corpuscular theory of light provided the rival theory with the wave mechanics its own champions had been unable to develop for themselves.'

Simple harmonic motion is defined in Book II, Proposition 47: 'If pulses are propagated through a fluid, the several parts of the fluid, going and returning with the shortest reciprocal motion, are always accelerated or retarded according to the law of the oscillating pendulum.' The Proposition shows by an extensive analysis, that the individual moving particles of a fluid undergoing wave motion oscillate in the same way as the bob attached to a simple pendulum. In the case of a gas, or 'elastic fluid', Boyle's law, with pressure proportional to density, ensures that a compression will produce a near linear reduction of the fluid particles' distance from the centre of vibration, while a rarefaction will produce the equivalent extension, leading to the $F \propto -r$ condition required for a harmonic oscillator. A Corollary states that 'the number of the pulses propagated is the same with the number of vibrations of the tremulous body, and is not multiplied in their progress.'

There had previously been a series of related propositions in Book I on isochronous motion and the cycloid. The significance of the cycloidal pendulum, of course, was that constraining the swing to a cycloidal arc ensured that the motion was isochronous whereas the semicircular arc of the simple pendulum motion ensured that it was not, and that the error increased with the angle of the swing. Book I, Propositions 48 and 49 concern rectifications of the general arcs of an epicycloid and a hypocycloid, by comparing limit-increments of the arc and the corresponding arc of the generating circle.[2] They generalise the rectification by Roberval of the simple cycloid in *De longitudine trochoidis*.[3] In Proposition 49, Newton, using a limit-increment argument, derives the 'isochronous oscillatory motion in a general force-field,' in 'the direct-distance case', where the tautochrone (or curve in which the time in which an object falls to its lowest point without resistance under uniform gravity is not determined by its starting point) is a hypocycloid.[4] Proposition 50 shows how 'to cause a pendulous body to oscillate in a given cycloid' by suspending it along the similar evolute cycloid. Proposition 51 demonstrates that, under a direct-distance force to the deferent centre, the constrained oscillations in a hypocycloid are isochronous. The dynamic condition for simple harmonic motion in an isochronous pendulum is that the component of force at a tangent to the path, and therefore effective in accelerating the bob, is proportional at every

---

[2]These occur in Section X, which incorporates Propositions 46–56.
[3]Roberval (c 1640/1693).
[4]First edition, 146–150, Whiteside, *MP* VI, 23.

point to the pendulum's displacement from the position of equilibrium. This was the first analysis of the phenomenon.

In Proposition 52, Newton finds mathematical equations relating such quantities as force, acceleration, velocity, displacement and time in the case of simple harmonic motion, establishing the velocity of the bob at any point in the motion, and the period for the entire oscillation. The second case considered compares the oscillations produced in unequal (hypo)cycloidal arcs. Simple harmonic motion requires the oscillations to be isochronous or periodically regular, and Proposition 53 defines the general condition that will produce isochronous oscillation for any given law of central force, 'granting the quadratures of curvilinear figures.' Proposition 54, again requiring quadrature, shows how to determine the time period for an isochronous oscillation using a geometric construction. Propositions 55 and 56 provide minimal generalisations, which analogously determine the planar oscillatory motion induced by a given force-field 'in any curve surperficies,' or spiral surface, 'whose axis passes through the centre of force.'

Newton, in Book II, Propositions 25–29, also considers 'damped' harmonic motion in a uniformly dense medium, based on the equation

$$ds^2/dt^2 = v \cdot \mathrm{d}v/\mathrm{d}s = -ks + lv^n, \quad n = 0, 1, 2,$$

and the assumption that the resistance ρ is constant, or proportional to the speed v, the square of the speed v^2, or a combination of these. Whiteside considers these propositions to be of considerable interest for the 'unique illumination they shed on Newton's mature power to formulate, and the limitations of his geometrical expertise accurately and adequately to substantiate, viable exact solutions to "infinitesimal" equations not in immediately quadrable form.'[5]

After establishing, in Proposition 24 and its seven Corollaries, that the length of a simple pendulum is proportional to the weight and the square of the time period, Newton shows, in Proposition 25, that 'Pendulous bodies that are, in any medium, resisted in the ratio of the moments of time, and pendulous bodies that move in a non-resisting medium of the same specific gravity, perform their oscillations in a cycloid in the same time, and describe proportional parts of arcs together.'[6] In the next Proposition 26, he sets

---

[5]Whiteside, *MP* VI, 439. Section VI of Book II, on resistance to pendulous bodies, incorporates Propositions 24–31.

[6]The more modern 'pendulous' replaces Motte's archaic word 'funependulous' here and elsewhere. The Corollary to this Proposition states that the swiftest motion is now no longer in the lowest place.

out to prove that a simple pendulum, resisted in the ratio of velocity, will be isochronous in the same way as a non-resisted pendulum. Though the demonstration is erroneous, the result is correct.[7] However, the error in the demonstration is carried over into Proposition 27, for a pendulum resisted in the ratio of velocity squared; and Whiteside asserts that its conclusion that the difference in time periods in a resisting medium and a non-resisting medium of the same density is 'nearly' proportional to the arcs described, is 'rawly approximate at best.'[8]

Proposition 28, however, returns to correct procedures in demonstrating that: 'If a pendulous body, oscillating in a cycloid, be resisted in the ratio of the moments of the time, its resistance will be to the force of gravity, as the excess of the arc described in the whole descent above the arc described in the subsequent ascent' is 'to twice the length of the pendulum.' Proposition 29 sets out to find the resistance at each place for a body oscillating in a cycloid which is resisted as the square of the velocity. Because this calculation is difficult, Newton adds Proposition 30, based on energy principles. 'Newton's stated equality is rigorously exact,' according to Whiteside,[9] and Mervyn Hobden has shown that it incorporates 'the first ever phase plane representation of an oscillating system,' allowing a purely geometrical analysis.[10] In a historical perspective, this was a startling innovation, inaugurating a style of dynamic representation which achieved perfection only with the phase diagrams introduced by Henri Poincaré at the end of the 19th century and featuring prominently at the present time in the mappings of canonical variables for chaotic systems.

Proposition 31, which says that an increase or decrease in resistance to 'the proportional parts of the arcs' in a given ratio will lead to the difference between the arc in descent and subsequent ascent being increased or decreased in the same ratio, was used as the basis for an experimental investigation of the resistances of air, water and other fluids, described in the General Scholium which follows. In principle, Newton was able to investigate

---

[7]Whiteside considers this lucky (*MP* VI, 442), but Newton's instinct for a correct result generally goes beyond fortunate coincidence.

[8]Whiteside, *MP* VI, 442. Newton's intentions, however, in this and other propositions in this section, may have been intentionally aimed at qualitative 'mechanical' or geometrical explanation rather than quantitative algebraic precision involving equations devoid of analytical solutions. It is clear from the wording that the proposition is considered to be largely qualitative and approximate. See Hobden, 2011.

[9]Whiteside, *MP* VI, 446.

[10]Hobden (2011). The abscissa is effectively displacement and the ordinate velocity. The diagram uses a half plane, so 'there are no closed integral curves.'

the viability of his model for a resistance force based on a variation with velocity of the form $a_1v + a_2v^2$, or any other power of v, by a curve-fitting process, which, to a large extent, showed that, under the conditions he was investigating, the variation was predominantly with v^2, although he had to include a variation with $v^{3/2}$ to produce a good enough fit with the data. He also found that experiments with conical pendulums gave general agreement with the total value of resistance, although the method was intrinsically less accurate. He discovered later that the motion of the pendulum bob created a to-and-fro motion in the resisting fluid, making the measured velocity of the bob different from its velocity with respect to the fluid, and he subsequently obtained better results on resistance forces using the simpler method of measuring the speeds of falling spheres.

While authorities such as Truesdell and Whiteside, have seen the analysis of resisted harmonic motion in Propositions 24–31 in terms of an equivalent algebraic analysis not used by Newton, often involving nonlinear equations, Hobden has shown that Newton's geometrical constructions were intended in a more practical way as 'moving pictures', dynamical rather than static 'representations of the system,' leading to approximate or semi-qualitative solutions that could be used in the most important cases. Newton, for example, knew that it was impossible to fix a point in space at which a real pendulum would come to rest, and Hobden considers his geometrical insights to be in a number of ways 'superior' to the algebraic analyses found in 'modern textbooks'. Proposition 31 and its five corollaries used energy considerations to show that a change in resistance, based on any power of velocity, would lead to a corresponding change in the pendulum arc, while the corollary to Proposition 30 used the phase plane diagram to calculate the work done against resistance and establish a reduction by a factor of $7/11$ or $2/\pi$ for a resistance proportional to velocity.[11]

For a resistance proportional to velocity squared, the corresponding factor was $3/4$, and for the hypothetical resistance proportional to $v^{3/2}$, the factor was $7/10$. Newton's experiments, which measured the decrease in amplitude due to resistance in a very long pendulum using different angles of release, were used to establish the relative contributions of the three components to the total resistance. They also showed that, for media that were not too viscous, the part of the resistance force proportional to the square of the velocity was also proportional to the density of the medium,

---

[11]Hobden (2011); see also Hobden (1981–1982); Harrison (1988). Hobden says that Propositions 30 and 31 are examples of a phase plane delta method.

being 13–14 times greater in mercury than in water, and 850 times greater in water than in air.[12]

Newton's analysis, which was aimed at establishing the validity of the experimental demonstration of the principle of equivalence, which he had described in Corollary 7 to Proposition 24, would have an extraordinary technological consequence within a few years of his death. John Harrison, in attempting to construct a chronometer to solve the problem of longitude, had already realised that a perfect harmonic motion could only be achieved using an anharmonic force transfer. To achieve the most accurate result, however, he appears to have referred to Newton's work on resisted harmonic motion and constructed his famous grasshopper escapement in such a way that the arc ratio between resisted and unresisted oscillations would be close to Newton's $2/\pi$, or, in approximate terms, $2/3$, and that a compensating additional force in the second half of each cycle would create an apparent harmonicity in the total motion.[13] Harrison's spectacular success came from a combination of his unrivalled mechanical genius and his realisation from theoretical arguments that the 'linear oscillator', as Leon Brillouin would later describe it, was 'an anomaly', unobtainable within any closed mechanical system.[14]

The definition of simple harmonic motion in Book II of the *Principia* leads on to that of wave motion. Proposition 43 describes how waves are propagated: 'Every tremulous body in an elastic medium propagates the motion of the pulses on every side right forward; but in a non-elastic medium excites a circular motion.'[15] Two cases are described. The first describes longitudinal compression waves. 'The parts of the tremulous body, alternately going and returning, do in going urge and drive before them those parts of the medium that lie nearest, and by that impulse compress and condense them; and in returning suffer those compressed parts to recede again, and expand themselves.' The manuscript *De aere et aethere* has a

---

[12] *Principia*, Book II, General Scholium following Proposition 31.

[13] Hobden (2011). Harrison's work on chronometers was begun in the late 1720s and achieved its first notable success with the completion of the first of his famous precision sea clocks (H1) in 1735. Hobden says that the force asymmetry is not discussed in Harrison's manuscript of 1730, but the '2:3 impulse' is introduced in his *Explanation of my Watch or Timekeeper for the Longitude* of 1764 and so was probably a later development, perhaps originating in a first viewing of the 1729 translation of the *Principia* on his visit to George Graham and Edmond Halley in London during 1730.

[14] Hobden (2011), citing Brillouin (1964). The self-taught Harrison had originally been introduced to Newtonian principles by reading a copy of the lectures of Nicholas Saunderson, the successor to Whiston as Lucasian Professor.

[15] Propositions 41–50 constitute Section VIII, on waves, in Book II.

description, in a single unfinished sentence, of sound in air as a compression wave of this kind, with successive condensations and rarefactions.[16]

Proposition 43 continues:

> And though the parts of the tremulous body go and return in some certain and determinate direction, yet the pulses propagated from thence through the medium will dilate themselves towards the sides...; and will be propagated on all sides from that tremulous body, as from a common centre, in surfaces nearly spherical and concentric. An example of this we have in waves excited by shaking a finger in water, which proceed not only forward and backward agreeably to the motion of the finger, but spread themselves in the manner of concentric circles all round the finger, and are propagated on every side. For the gravity of water supplies the place of elastic force.

In the second case, if the medium is not elastic, the motion is propagated in an instant to the parts which yield most easily, generating a circular motion to the parts which the vibrating body by its motion would otherwise have left empty, as happens with a projectile.

Proposition 44 relates the motion of a liquid column in a U-shaped canal, which Newton sees as analogous to wave motion, to that of a simple pendulum. The characteristics of wave motion can thus be derived by analogy to those of the simple harmonic oscillator. Corollary 1 says the frequency is independent of the amplitude. Corollary 2 says that the water will descend in 1 second and ascend again in 1 second, if the canal is 6 1/9 French feet, for this is twice the length of a seconds pendulum. Corollary 3 says the time period is proportional to the square root of the length of the canal. Propositions 45 and 46 say this is also true of the velocity of the waves, which (in an incompressible medium) is consequently proportional to the square root of the wavelength.[17] The discussion leads on to the definitions of frequency and wavelength, and the fundamental relation between them and the velocity of the waves:

$$\text{velocity} = \text{frequency} \times \text{wavelength}$$

or

$$v = \nu\lambda.$$

---

[16] *De Aere et Aethere*, *Unp.*, 217–218, translation, 224–225.
[17] Propositions 45, 46 and Corollaries 1, 2. In the equivalent modern formula $c^2 = g\lambda/2\pi$ (Pask, 2013, 385).

The definition of wavelength is given in Proposition 46: 'That which I call the breadth of the waves is the transverse measure lying between the deepest parts of the hollows, or the tops of the ridges.' Newton states a verbal equivalent of the relation

$$\text{period of oscillation} \times \text{velocity} = \text{wavelength}$$

or

$$(2\pi/\nu) \times \sqrt{gh} = \lambda$$

at the end of this proposition, without any calculation.[18] In Corollary 1, he says that waves of wavelength 3 1/18 French feet (the length of a seconds pendulum) will have a time period of 1 second. As Newton says, the expression for the time period in Proposition 46, based on infinitesimal amplitudes, is approximate compared to the exact value assumed in Proposition 45.[19]

The definition of frequency follows in the Corollary to Proposition 47: '...the number of the pulses propagated is the same with the number of vibrations of the tremulous body, and is not multiplied in their progress.' The relation of these to velocity is given in Proposition 50 with a direct statement of $c/\nu = \lambda$: 'Let the number of vibrations of the body, by whose tremor the pulses are produced, be found to any given time. By that number divide the space which a pulse can go over in the same time, and the part found will be the breadth of one pulse.' Wave numbers or spatial frequencies (reciprocals of wavelengths) are calculated, for the first time in physics, in Newton's work on thick optical plates.[20] Significantly, $v = \nu\lambda$ and several other wave formulae are attributed to light as well as to sound. Direct comparisons of its application to light and sound had been made as early as the *Hypothesis of Light* of 1675.[21]

In the first edition of the *Principia*, Newton referred to experiments made by Mersenne in Paris in which a stretched musical string vibrating 104 times a second made a sound of the same pitch as an open organ pipe four foot in length and a stopped organ pipe two foot in length. With his relation between velocity, frequency and wavelength, Newton was able to give

---

[18] Chandrasekhar (1995, 585).

[19] Chandrasekhar (1995, 585).

[20] *Opticks*, II, Part 4, Observation 8. Thick plates produce analogous ring phenomena to thin films, but the analogy is 'not actually close', and reverses the light and dark ring conditions, or those of transmission and reflection. However, by 'using a concept of path difference,' Newton 'was able to transform the problem of thick plates to an equivalent one of thin films and calculate the diameters of the rings' (Shapiro, 2013, 190).

[21] *Hyp.*, *Corr.* I, 361–386, 366.

the wavelength as '$9^{1/4}$ feet, that is, roughly twice the length of the pipe. Hence, it is likely that the lengths of the pulses in the sounds of all open pipes are equal to twice the lengths of the pipes.'[22] The later experiments of Joseph Sauveur, with a frequency of 100 pulses per second, corresponding to the pitch of an open pipe of approximately five Paris feet, and the superior measurement of the velocity of sound, by William Derham (1705), of 1,142 London feet or 1,070 Paris feet per second, made this conclusion increasingly probable.[23]

In the later editions of *Principia*, Newton concludes again, from the new data, 'it is probable that the breadths of the pulses, in all sounds made in open pipes, are equal to twice the length of the pipes.'[24] This is a significant observation in its own right; it happens because the pipes create standing waves by reflection of the incident wave and so the fundamental mode measures half a wavelength — in closed pipes (pipes closed at one end), as would have been apparent from the original observations, it is only a quarter of a wavelength. Sound amplification by megaphones and the relatively slow attenuation of sound amplitude in tubes that impede the expansion are explained by such reflection at the end of the same Scholium.

Proposition 46 describes the propagation of long waves of minimal amplitude in canals. Chandrasekhar believed that Newton, in this analysis, had effectively 'discovered the Lagrangian displacement,' ζ, where the horizontal velocity $u = \partial\zeta/\partial t$. Chandrasekhar's analysis requires the wave equation, a result first published in the mid-18th century, and it would be interesting to know if this was a necessary step in Newton's derivation.[25] Without direct evidence this must remain extremely problematic. Proposition 47, as we have seen, shows how the particles of a fluid in undulatory motion are necessarily acting as harmonic oscillators with the same motion as the bob of a simple pendulum.[26] Newton also makes the analogy here, and in Proposition 49 and its corollaries, between the wave motion and a radius in circular motion projected onto a fixed diameter.

Another very important wave formula due to Newton, and derived from cases of simple harmonic motion, is that relating the velocity (c) to the

---

[22]II, Proposition 50, Scholium, first edition, Cohen and Whitman (1999, 777).

[23]Sauveur (1700) and Derham (1708–1709).

[24]II, Proposition 50, Scholium.

[25]Chandrasekhar (1995, 585–586).

[26]II, Proposition 47 and Corollary.

elasticity (k) and density (ρ) of the medium

$$c = \sqrt{\frac{k}{\rho}}.$$

Proposition 48 states that: 'The velocities of pulses propagated in an elastic fluid are in a ratio compounded of the square root of the ratio of the elastic force directly, and the square root of the ratio of the density inversely; supposing the elastic force of the fluid to be proportional to its condensation.' The formula shows that the velocity is independent of both the amplitude of vibration and the frequency or pitch. Proposition 49, which is aimed at finding the velocity of the pulses, given the density and elastic force of the medium, is included to clarify the demonstration in the previous proposition.

A crucial component of such later aether theories as those of Fresnel and Maxwell, this was again applied by Newton to both light and sound. 'The last Propositions respect the motions of light and sounds; for since light is propagated in right lines, it is certain that it cannot consist in action alone. As to sounds, since they arise from tremulous bodies, they can be nothing else but pulses of the air propagated through it...'[27]

Newton used the formula to calculate the velocity of sound at 968 feet per second in the first edition of the *Principia*. At first this agreed well with experimental results made by Newton himself, but later results suggested a discrepancy of about one-sixth between theory and experiment. Laplace brilliantly solved the problem by exchanging the isothermal value for the elasticity of air used by Newton with the adiabatic value which was 1.4 times greater and which Newton could not have known about. Despite supplying this correction, Laplace nevertheless considered the analysis 'a monument to Newton's genius' as the formula is substantially correct.[28]

Newton's measurements of the velocity of sound were also among the earliest realistic ones. Earlier in the century, Giles Roberval had found a value of 600 feet per second, which was far too low, and Marin Mersenne had found 1,474 feet per second, which was somewhat too high, though he also found a value, using a different type of measurement, of 1,036 feet per second.[29] Newton, in his earlier experiments, found that an echo travelled 416 feet in the cloisters of Neville's Court at Trinity College in a time that was slower than the swing of a pendulum 5.5 inches long but was faster than that of a pendulum 8 inches long, making the velocity between 920 and 1,085

---

[27] II, Proposition 50, Scholium.
[28] Laplace (1816).
[29] Mersenne (1636, 44); Mersenne (1644, 140) (for Roberval).

feet per second. This may be compared with the value of about 1,126 feet per second at 20°C in dry air, which reduces to about 1,087 feet per second at 0°C. Newton's later experiments, of 1694, placed the limiting values at 984 and 1,109 feet per second.[30] On 28 February 1705, William Derham timed the interval between the flash and roar of a cannon at Blackheath from his church tower at Upminster 12 miles away, and calculated the velocity at 1,142 feet per second for the paper he published in the *Philosophical Transactions* of 1708.[31]

The failure to obtain an exact match between theory and experiment troubled Newton, as it was inconceivable to him that the formula could be incorrect. He correctly guessed that he had underestimated the constant of elasticity, but was unable to explain why. He tried arguing that air molecules must take up some finite space ('crassitude'). The presence of water vapour might also affect the result. If the spacing of the air molecule was about 9.5 times those of water, the density ratio would be 1/850. Scholars, including Westfall, formerly classed this as a subterfuge, on a par with the lunar test and precession,[32] but, as with those cases, the criticism is unjustified, though this time for a different reason. In the lunar test, Newton simply stated more figures than the accuracy of the measurement would allow (a relatively common practice up to the 19th century, but of no great significance in the context). In the calculation of precession, he chose the figures from the available range which would give the most desired result (a practice still followed on occasions for speculative work), but didn't claim that he had done anything else. In the case of the velocity of sound, Newton stated clearly that he had found a discrepancy, and made one or two suggestions about how it might be overcome, but leaving these open to subsequent investigation.[33]

---

[30]Westfall (1980, 734).

[31]Derham (1708–1709).

[32]Westfall (1980, 734–736). Truesdell (1967, 201), very often a harsh critic of Newton (no doubt partly because of his desire to stress the importance of Euler), was the first to introduce the idea of a 'fudge' factor for this calculation, and hence in Newton's work generally. It was taken over by Westfall and hence by many others who simply, and uncritically, quoted Westfall as an authority, but it is the least convincing of all three cases. If this is a 'fudge' then all of science which adopts an iterative theoretical procedure for dealing with discrepancies between theory and experiment is a fudge!

[33]Cohen (1999, 361–362), has stated quite clearly that he believes the charge to be an error in this case. It is interesting that judgements have been made on the basis of the standards that Newton himself created. Prior to his work, the *whole* explanation would have been along the lines that he was now using to suggest a possible way of accounting for a 16% discrepancy!

Newton's attempts to explain the discrepancy numerically are in fact an interesting demonstration of the difference between abstract analytical and hypothetical reasoning. His original formula was derived from the former and is still valid if we use the correct constant of elasticity. His attempts to show the reason for the discrepancy are the result of making up arbitrary mechanical hypotheses, and show how success using this methodology is spectacularly unlikely. Even Newton has little aptitude for it. He would certainly have avoided it altogether if he hadn't felt under pressure to come up with a result that made a match between theory and observation. He was well aware that such matches are not always achieved at the first attempt even if the theory is basically correct.

3.2. Interference and Diffraction

Newton famously introduced a principle of interference in discussing the tides at Batsham, a port situated in the Gulf of Tonkin on the coast of what is now Vietnam, which had been visited by a mariner named Francis Davenport, whose eye-witness account had been discussed in a paper by Halley. The discussion was used by Young when he was trying to establish the principle of interference between two light beams, though there is no reason to suppose that Newton ever thought of applying the principle to light waves.[34]

> The two luminaries [Sun and Moon] excite two motions which will not appear distinctly, but between them will arise one mixed motion compounded out of both.... Further, it may happen that ...the same tide, divided into two or more succeeding one another, may compound new motions of different kinds... An example of all of which Dr. *Halley* has given us, from the observations of sea men in the port of *Batsham*, in the kingdom of *Tunquin*... In that port, on the day which follows after the passage of the Moon over the equator, the waters stagnate: when the Moon declines to the north, they begin to flow and ebb, not twice as in other ports, but once only every day: and the flood happens at the setting, and the greatest ebb at the rising of the Moon.[35]

The principle of interference requires a complete understanding of phase, with constructive interference for waves in phase, and destructive interference for waves in antiphase, and this occurs elsewhere in Newton's work, for example, in the theory of fits and in the case of sound waves, where he finds that notes of the same pitch or frequency, will not necessarily be in phase when they reach the ear if they have to travel different path lengths.

---

[34]Halley (1684); Young (1802a).
[35]III, Proposition 24.

In a letter of 1677, Newton explained that two sources with the same pitch would not produce a harmonised sound to a listener, because they would not necessarily be in phase at the ear. Rather, he said, 'unisons are rather a harmony of two like tones then a single tone made more loud and full by the addition.'[36] This is a description of the phenomenon of beats or amplitude modulation.

Propositions 41 and 42 of Book II of the *Principia*, with accompanying diagrams, show that Newton also understood wave diffraction, treated in terms of the secondary wavelet principle. Here, he writes: 'A pressure is not propagated through a fluid in rectilinear directions unless where the particles of the fluid lie in a right line.' 'All motion propagated through a fluid diverges from a rectilinear progress into the unmoved spaces.' Newton proved Proposition 42 by an argument which is not too different from the one that Huygens exploited for light waves (though Huygens rejected the idea of diffraction), by showing how a pressure wave in a particulate fluid medium produces secondary wavelets which spread out through a small aperture in a solid barrier, creating a new wavefront on the other side.[37]

Rectilinear motion, however, seemed to rule this out for light, which could not be described by pressure waves like those for sound or those generated in water.

> If Light ... consisted in Pression or Motion, propagated either in an instant or in time, it would bend into the Shadow. For Pression or Motion cannot be propagated in a Fluid in right Lines, beyond an Obstacle which stops part of the Motion, but will bend and spread every way into the quiescent Medium which lies beyond the Obstacle.... The Waves on the Surface of stagnating Water, passing by the sides of a broad Obstacle which stops part of them, bend afterwards and dilate themselves into the quiet Water behind the Obstacle. The Waves, Pulses or Vibrations of the Air, wherein Sounds consist, bend manifestly, though not so much as the Waves of Water.... But Light is never known to follow crooked Passages nor to bend into the Shadow. For the fix'd Stars by the Interposition of any of the Planets cease to be seen. And so do the Parts of the Sun by the interposition of the Moon, *Mercury* or *Venus*. The Rays which pass very near to the edges of any Body, are bent a little by the action of the Body...; but this bending is not towards but from the Shadow, and is perform'd only in the passage of the Ray by the Body, and at a very small distance from it. So soon as the Ray is past the Body, it goes right on.[38]

---

[36] Newton to Francis North, 21 April 1677, *Corr.* II, 205–207.
[37] Stuewer (1970).
[38] Q 28.

It is interesting that he actually considered the possibility of light bending into the shadow, even if only to reject it, for his contemporary Christiaan Huygens, who had a wave theory of light based on random pulses, thought that the effects that we now know as light diffraction had no experimental validity at all.[39]

Newton tried hard to understand the diffraction effects observed by Francesco Grimaldi, a Jesuit Professor in Bologna, in the shadows of objects, probably without first-hand knowledge of Grimaldi's own descriptions — he read about them in Honoré Fabri's *Dialogi Physici* (1669), which he received as a present from John Collins in May 1672.[40] Grimaldi saw that light was bent in passing the sides of objects, thus extending shadows and producing coloured fringes at their edges. Newton was seriously interested in the implications of Grimaldi's work, and tried to repeat his experiments showing internal fringes. His quantitative experiments were the best before the 19th century.

> Grimaldo has informed us [Newton began], that if a beam of the Sun's Light be let into a dark Room through a very small Hole, the shadows of things in this Light will be larger than they ought to be if the rays went on by the Bodies in streight Lines, and that these shadows have three parallel fringes, bands or ranks of coloured Light adjacent to them. But if the Hole be enlarged the fringes grow broad and run into one another, so that they cannot be distinguished.[41]

Careful experiments on diffraction (or, as he called it, 'inflection') from straight and circular edges, both transparent and opaque, revealed 'exterior' coloured fringes, bending away from the geometrical shadow, and generally following the spectral sequence from violet to red.

> Now since this inflection of the rays is performed in the air without the knife [the diffracting object], it follows that the rays which fall upon the knife are first inflected in the air before they touch the knife. And the case is the same of the rays falling upon glass. The refraction, therefore, is made not in the point of incidence, but gradually, by a continual inflection of the rays: which is done partly in the air before they touch the glass, partly (if I mistake not) within the glass after they have entered it . . .[42]

---

[39] Huygens (1690/1912); Shapiro (1974, 224).

[40] Grimaldi (1665); Fabri (1669). Newton knew about optical diffraction before Hooke, who first reported on it as a 'new' effect in 1675. He had even referred to what we know to be diffraction effects in the *Quaestiones Quaedam Philosophicae* of the 1660s (see p. 75 and n. 51).

[41] *Opticks*, Book III, Part 1.

[42] *Principia*, I, Proposition 96, Scholium.

Similar effects were produced by obstacles that were long and thin such as human hairs, threads, pins and straws.[43] Using other media showed that it was not due to 'the ordinary refraction of the Air,' as had 'been reckoned by some.' Newton measured the fringe separations at different distances between object and screen, finding that the fringe spacings and also the breadth of the shadow were not altered by increasing this distance.

According to Observation 5, he saw sunlight bend into the shadow of a knife edge. 'I let part of the Light which passed by, fall on a white Paper two or three Feet beyond the Knife, and there saw two streams of faint Light shoot out both ways from the beam of Light into the shadow like the tails of Comets.' Attempting to discover the origin of this light, Newton viewed the edge of the knife when looking against the sunlight incident on it. He observed a line of light on its edge. 'This line of Light appeared contiguous to the edge of the Knife, and was narrower than the Light of the innermost fringe, and narrowest when my Eye was furthest from the direct Light, and therefore seemed to pass between the Light of that fringe and the edge of the Knife, and that which passed nearest the edge to be most bent, though not all of it.'[44]

Single slit diffraction (between two knife edges) showed a pattern with a bright line at the centre and three coloured fringes on each side. Narrowing the slit widened the pattern. At a separation of 1/400 of an inch the pattern disappeared.

> And when the distance of their edges was about the 400th part of an Inch the stream parted in the middle, and left a Shadow between the two parts. This Shadow was so black and dark that all the Light which passed between the Knives seem'd to be bent, and turn'd aside to the one hand or to the other. And as the Knives still approach'd one another the Shadow grew broader, and the streams shorter at their inward ends which were next the Shadow, until upon the contact of the Knives the whole Light vanished leaving its place to the Shadow.[45]

Newton also observed that diffraction fringes were generated when light was directed through a very narrow glass wedge.[46]

Though he frequently refers to the production of 'three' coloured fringes, Newton was aware that the number of fringes was not limited in this way for he found four or five fringes from 'Plates of Looking-glass sloop'd off

---

[43] *Opticks*, Book III, Part 1, Observations 1, 3 and 4.
[44] Observation 5.
[45] Observation 5 (1704 edition, 122).
[46] Observation 1 (1704 edition, 116–117).

near the edges with a Diamond-cut.'[47] A curious experiment from the series on prisms showed the dispersed light from a prism passing through a fine comb, with a spacing of seven teeth and seven interstices of equal width per inch, creating a set of parallel spectra on a paper placed two or three inches away, each of the intervals between the teeth of the comb 'producing the Phænomena of one Prism.' Extending the distance to a foot or a little more extended the spectra so that they merged into each other and became white light. Obstructing one of the interstices allowed the neighbouring spectra to extend into the space and show colour again.[48]

Other experiments involving multiple apertures clearly involved diffraction. An experiment recorded in the diffraction section of the *Opticks*, which required 'looking on the Sun through a Feather or black Ribband held close to the Eye,' disclosed 'several Rainbows', 'the Shadows which the Fibres or Threds cast on the *Tunica Retina*, being border'd' with the characteristic coloured diffraction fringes.[49] This observation has been connected with one made by James Gregory in correspondence with John Collins on 13 May 1673, which he hoped would be commented on by Newton:

> If ye think fit, ye may signifie to Mr Newton an small experiment, which (if he know it not alreadie) may be worthy of his consideration. Let in the Sun's Light by a small hole to a darkened house, and at the hole place a feather (the more delicate & white, the better for this purpose) and it sall direct to a white wall or paper opposite to it a number of small circles and ovales, (if I mistake them not) whereof one is someqt white (to wit the midle, which is opposite to the Sune) and all the rest severallie coloured: I wold gladlie hear his thoughts of it.[50]

We do not know whether Newton received information of Gregory's suggestion, either from Collins or from Gregory himself if he called in at Cambridge on his return journey from London to Scotland in September 1673. However, Newton's experiment is very different from Gregory's and

---

[47] Observation 3.

[48] *Opticks*, Book I, Part 2, Proposition 5, Experiments 12–13. An earlier version had used slits cut into a sheet of 'paper or other thin opaque body,' which was placed variously on the side of the incident and of the refracted light (*Lectiones Opticae*, Lecture 4, 43–44, *OP* I, 102–105). An even earlier version, with just four slits, can be found in the *Essay of Colours*, §47, CUL Add 3075, ff. 7ᵛ–8ʳ. The experiment is to be distinguished from that made with a much coarser comb (*Opticks*, Book I, Part I, Proposition 5, Experiment 10), where the much larger spacings served to separate out the colours of a single spectrum into something close to monochromatic light and to combine them by motion into an overall whiteness.

[49] *Opticks*, Book III, Part 1, Observation 2.

[50] *Corr.* I, 378–380.

much less sophisticated, being concerned with direct observation rather than projection. It is almost impossible to believe that, with Gregory's experiment available to him, he would have advocated direct observation of the Sun when his early experiment on looking at the *reflection* of the Sun had nearly caused permanent damage to his eyesight. The experiment seems, rather, even to the point of using the same wording, to be the one recorded in *Quaestiones Quaedam Philosophicae* of c 1664–1666 in which 'A feather or black ribband put twixt my eye & the setting sunne makes glorious colours.'[51] Two things are specially worth noting here. The phrase 'glorious colours' is the most aesthetically responsive Newton ever used in relation to any of his optical experiments and is exactly what might be expected of multiple overlapping spectra. In addition, it seems that Newton, in apparently referring to one of his own experiments, can be completely exonerated from any charge of appropriating the results of a contemporary without acknowledgement. It seems likely that Newton, in not adopting what would have been a neat extension of his own work on prismatic spectra, had never heard of Gregory's suggestion, and that this can be attributed to his reluctance at the time to communicate with John Collins.

These various observations by Newton and Gregory on multiple spectra from feathers and ribbons can, at any rate, be regarded as ancestral to those of the eighteenth and nineteenth centuries using diffraction gratings. They are also, along with Newton's experiments using combs, the first experiments to investigate the optical effects derived from multiple slits, an issue which later became significant in the interpretation of quantum mechanics.

Another significant wave concept is decoherence. In discussing telescopes in the *Opticks*, Newton writes:

> For the Rays of Light which pass through divers parts of the aperture, tremble each of them apart, and by means of their various and sometimes contrary Tremors, fall at one and the same time upon different points in the bottom of the Eye, and their trembling Motions are too quick and confused to be perceived severally. And all these illuminated Points constitute one broad lucid Point, composed of those many trembling Points confusedly and insensibly mixed with one another by very short and swift Tremors, and thereby cause the Star to appear broader than it is, and without any trembling of the whole.[52]

The 'tremors' described here are those of air molecules, and Newton is using them to explain atmospheric refraction in the manner initiated

---

[51] QQP, *Of Colours*, f. 133^v.
[52] *Opticks*, Book I, Part 1, Proposition 8.

by Hooke in *Micrographia*.[53] Both Newton and Hooke tended to subsume both diffraction and atmospheric refraction under the general heading of 'inflection'; it is interesting, therefore, to note that a substitution of intrinsic wavelike properties for the atmospherically-induced tremors would make this passage into a perfect qualitative explanation of the formation of an image by diffraction.

3.3. Periodicity and Newton's Rings

Following Lord Brouncker's publication of *Musica Compendium* in 1653,[54] Newton applied logarithms to the division of the musical scale, and was the first to use 'logarithmic notation' to represent 'the magnitude of intervals.' He anticipated 'the modern cent system' by taking 'the equal-tempered half tone as his basic unit or 'common measure', using this system' to find 'the ratios of the syntonic diatonic scale' or 'just scale'.[55] He also considered dividing the octave in other ways, for example, into 12, 20, 24, 25, 29, 36, 41, 51, 53, 100, 120 or 612 parts, of which he concluded that the one into 53 parts was the best. He devised a 'catalogue' of the '12 musical modes in their order of gratefulness' and created a system for switching between the different modes.[56] Newton's *Origins of Gentile Theology* and his correspondence with the Oxford undergraduate John Harington in 1698 show that he was interested in the significance of the kind of simple geometrical ratios that generated artistic and musical harmony for physics and cosmology as evidence for 'the wisdom or power of God.'[57]

Newton subsequently divided the optical spectrum into seven colours, by analogy with the seven tones of the musical scale. Originally he had five colours — red, yellow, green, blue and violet — but these were quickly supplemented by orange and indigo.[58] The analogy is actually quite good, though Newton was fully aware that an equal-tempered scale would fit the colour distribution as well as the musical scale, and the distribution cannot be fitted equally to all refracting substances. However, just as the frequencies

---

[53]Hooke (1665, 217–222, 228–240).

[54]Brouncker (1653).

[55]'Newton, Sir Isaac,' Grove Music Online.

[56]CUL, Add 3958.2, 34v/35r/37r; Add 4000, 104r-113r, 137v-143r. *MP* I, 298 ff, 219 ff and 369 ff. Gouk (1988, 1999).

[57]*Origins of Gentile Theology*, 1680s–1690s, NP. Harington to Newton, 22 May 1698, *Corr.* IV, 273–274. Newton to Harington, 30 May 1698, *Corr.* IV, 274–275. This is one of those correspondences that show the more positive side of Newton's character.

[58]*Lectiones Opticae*, Lecture 1, 2, *OP* I, 48–51 (five colours); *Optica*, II, Lecture 11, 98–99, *OP* I, 542–557 (musical scale and equal-tempered scale).

in the musical scale double over an octave, so the frequencies of visible light almost double in going from the extreme red $(4 \times 10^{14}\,\text{Hz})$ to the extreme violet $(7.7 \times 10^{14}\,\text{Hz})$. Also, just as there are two half-tones in the musical scale (E–F and B–C), so there are also two 'half-colours' (orange and indigo) in the visible spectrum. In fact, many people are unable to separate indigo from blue and and violet, and it is frequently asserted that Newton effectively 'invented' the colour to match the scale to the spectrum.[59] However, it now seems more likely that what Newton called 'blue' is what we call 'cyan', and that Newton's 'indigo' is our 'blue'. Newton actually refers to cyan (*cyaneum*) in *Optica*, saying that it can be made from sea green and indigo. He also indicates that he is aware that different individuals have different physiological responses to various colours, as the colour discrimination of his friend and assistant, John Wickins, is clearly superior to that of his own. So, Newton's spectrum of red, orange, yellow, green, blue, indigo, violet would be equivalent to our red, orange, yellow, green, cyan, blue, and violet.[60]

According to Shapiro, the division was made using a 'partial dispersion law' in which the range of each colour would be a fixed fraction of the spectrum's length, irrespective of the actual length or the dispersing medium, in line with his view that the colours of light rays were a fundamental aspect of the nature of light. In this way, a mathematical division of the spectrum would allow a prediction of the frequency associated with each colour, which, due to the considerable expanse of each of the colours, would fall within the observed range. The division, taken to the limit, led to an erroneous linear dispersion law, but Newton's reasoning, as analysed by Shapiro, has a particularly interesting structure. Here we imagine the visible spectrum being expanded continuously down to rays with zero refraction beyond the red, as Newton did in the *Optical Lectures*. We now imagine, again as Newton did, the length corresponding to the visible proportion of the spectrum as proportional to $R = \sin r$, and that of the entire spectrum, visible and invisible, as corresponding to $R - I = \sin r - \sin i$, with I treated as a constant for an incident beam of white light. This is equivalent to taking small-angle approximations $(R \sim r$ and $I \sim i)$, and, according to Newton, $R/(R - I)$ will be a constant.[61] The linear dispersion law then says that the

---

[59] *Optica*, II, Lecture 11, 98–101, *OP* I, 542–549. *Hyp.*, *Corr.* I, 376. *Opticks*, Book I, Part 2, Propositions 3 and 6; Book II, Part 1, Observation 14, and Book II, Part 4, Observation 8.
[60] Waldman (2002, 193). *Optica*, Part II, Lecture 8, 70, *OP* I, 506–507 (on cyan). *Hyp.*, *Corr.* I: 376, *Opticks*, Book I, Part II, Proposition 3, Experiment 7 (on colour discrimination). This would remove another conspiracy theory relating to Newton!
[61] Shapiro (1979, 111; 2005, 113).

ratio of the dispersion ΔR (or the difference in R between the two ends of the spectrum) and $(R - I)$ is a constant for all media.

Newton's reasoning connects strangely with a bizarre correspondence he is alleged to have had some years later with Joshua Maddock, a country doctor from Whitchurch, who sent him some essays on a new topic in optics, based on rays of darkness. The evidence we have comes from a letter by Newton of 7 February 1679, which Westfall regarded as a forgery, but which Hall treated as authentic. If authentic, it would make a more satisfying conclusion to the correspondence triggered by the report of 1672 than the last letter Newton sent to the Jesuit Anthony Lucas in 1678. Maddock seems to have suggested that dark rays might exist and be governed by the same law of refraction as light rays. Newton says that it is indeed an interesting question whether such rays exist and whether they would be subject to the ordinary laws of refraction or some other. He grants that Maddock has supplied skilful and 'exceedingly subtle' arguments and ingenious proofs, which, if true, would certainly advance the science of optics, if the lack of experimental evidence could somehow be made up. The tone is the one of extreme politeness such as Newton often used with correspondents who asked for his opinion or help on scientific matters — the expression 'your usual kindness' implies that this wasn't the first time they had corresponded. Newton, perhaps for once, uses a little irony in saying that he is at a loss to know how to think about an experimental verification, the Emperor Tiberius (who, according to Suetonius, could see at night and in the dark, though only for a short time) not being available to help him.[62]

The real irony, of course, is that 'invisible' or 'dark' rays, such as Newton assumed for mathematical purposes and Maddock apparently for physical, do exist, at both ends of the spectrum, and were involved in experiments that Newton later organised on radiant heat. The fact that the Newtonian solar spectrum extended beyond the visible was first established when Herschel, in 1800, used a black-bulb thermometer to show that the heating effect extended beyond the red region of the visible spectrum and was caused by what he called 'invisible light', and which we now call infra-red radiation. Herschel also showed that radiant heat from terrestrial sources had the same character as that from the solar spectrum and could be reflected and refracted according to the same laws as visible light and dispersed by a prism.[63]

---

[62] *Corr.* II, 287–288.
[63] Herschel (1800).

The musical analogy suggests how serious Newton was about the wave effects that he associated with light, even though he never ascribed to the 'hypothesis' that light was *actually* a wave motion. The most important of these effects was periodicity. One of the most significant experiments ever undertaken in optical physics is the one now known as 'Newton's rings', an investigation of a thin film effect, or a pattern produced in the air film between a convex lens and a glass plate. It is significant because it is the one which first established periodicity as a property of light. The musical scale analogy here becomes especially important because it is meaningless without the concept of periodicity.

While it is usually stated that Newton first got the idea of Newton's rings from the thin film effects described in Hooke's *Micrographia*, themselves stimulated by experiments reported by Boyle, Newton says in the *Opticks* that he first saw the rings by rotating a pair of prisms bound together in a parallelepiped, before he used the combination of lens and glass plate, the circular symmetry produced by the lens allowing a massive simplicity in both experiment and calculation.[64] The name 'Newton's rings' does not, in fact, refer to the phenomenon of interference effects in thin films, a discovery which Newton made no attempt to claim for himself,[65] but to the experiment that he devised to demonstrate that light had a mathematically measurable periodicity.

Newton's own descriptions of the experiment cannot be bettered: 'It has been observed by others (i.e. Boyle and Hooke), that transparent Substances, as Glass, Water, Air, &c. when made very thin by being blown into Bubbles, or otherwise formed into Plates, do exhibit various Colours according to their various thinness, altho' at a greater thickness they appear very clear and colourless.'[66] 'By looking through ... two contiguous Object-glasses, I found that the interjacent Air exhibited Rings of Colours, as well by transmitting Light as by reflecting it.'[67] '...in a dark Room, by viewing these Rings through a Prism, by reflexion of the several prismatick Colours...'[68] 'I have seen the lucid circles appear to be about twenty in number with as many dark ones between them, the colour of the lucid ones being that of the light,

---

[64] *Opticks*, Book II, Part 1, Observations 1–3. Boyle (1663, 1665), Observation 9, 47 f.
[65] The next quotation makes this clear. He doesn't attribute the phenomenon to Hooke because Boyle had reported it first.
[66] *Opticks*, Book II, Part 1.
[67] *Opticks*, Book II, Part 1, Observation 9.
[68] *Opticks*, Book II, Part 2.

with which the glasses were illustrated.'[69] '...the circles made by illustrating the glasses, for instance, with red light, appear red and black...'[70]

He found that the dark and coloured circles occupied opposite spaces for reflection and refraction: 'And by projecting the prismatick Colours immediately upon the Glasses, I found that Light which fell upon the dark Spaces which were between the Colour'd Rings was transmitted through the Glasses without any variation of Colour.'[71]

> And if the glasses were held between the eye and the Prismatic colours, cast on a sheet of white paper, or if any Prismatic colour was directly trajected through the glasses to a sheet of paper placed a little way behind, there would appear such other Rings of colour & darknesse (in the first case, between the glasses, in the second, on the paper,) oppositely corresponding to those which appeared by reflexion: I meane, that, whereas by *reflected* light there appeared a black spot in the midle, & then a coloured circle; on the contrary by *transmitted* light there appeared a coloured spot in the midle, and then a black circle; & so on; the diameters of the coloured Circles, made by transmission, equalling the diameters of the black ones made by reflexion.[72]

The colours produced by the reflection and refraction process were what we now call complementary ones, adding up to produce white light. Intriguingly, the rings display alternate reflection and transmission in what we might now recognise as our primary and secondary colours (if we take Newton's 'blue' as our cyan — his diagram shows red opposite 'greenish blue' and 'bluish green').

> Comparing the colour'd Rings made by Reflexion, with those made by transmission of the Light; I found that white was opposite to black, red to blue, yellow to violet, and green to a Compound of red and violet [our magenta]. That is, those parts of the Glass were black when looked through, which when looked upon appeared white, and on the contrary. And so those which in one case exhibited blue, did in the other case exhibit red. And the like of the other Colours.[73]

If we take 'blue' to mean 'cyan', then these are exactly the complementary primary and secondary colours we recognise today: red/cyan, green/magenta, and violet/yellow. Elsewhere in the *Opticks*, Newton drew

---

[69] *Hyp.*, *Corr.* I, 380.

[70] *Hyp.*, *Corr.* I.

[71] *Opticks*, Book II, Part 1, Observation 15.

[72] *Hyp.*, *Corr.* I, 380.

[73] *Opticks*, Book II, Part 1, Observation 9.

attention to an experiment by Hooke with red and blue liquids, which were individually transparent, but taken together would pass no light through,[74] which he could now explain as the result of each liquid selectively absorbing the rays that the other would let through.

Newton applied the concept of wavelength to the sequence of colours: '...according to those Observations the thickness of the thinned Air, which between two Glasses exhibited the most luminous parts of the first six Rings were 1/178000, 3/178000, 5/178000, 7/178000, 9/178000, 11/178000 parts of an Inch.'[75] '...their arithmetical Means 2/178000, 4/178000, 6/178000, &c. being its Thicknesses at the darkest Parts of all the dark ones.'[76] '...I found the Circles which the red Light made to be manifestly bigger than those which were made by the blue and violet... By the most of my Observations it was as 14 to 9... And it was very pleasant to see them gradually swell or contract accordingly as the Colour of the Light was changed.'[77] In an earlier account, he said: '...in plates of denser transparent bodyes the rings are made at a lesse thicknesse of the plate, (the vibrations I suppose being shorter in rarer æther then in denser).'[78]

Periodicity in the rings was always associated with the size of vibration, whatever that may be: 'And from thence the origin of these Rings is manifest; namely, that the Air between the Glasses, according to its various thickness, is disposed in some places to reflect, and in others to transmit the Light of any one Colour ...'[79] '...colours, like sounds, being various according to the various bignesse of the Pulses.'[80] '...the Arithmeticall progression of those thicknesses, wch reflect & transmit the rayes alternately, argues that it depends upon the number of vibrations between the two Superficies of the plate whether the ray shall be reflected or transmitted...'[81]

For compound light, the colours overlapped because of the varying wavelengths:

> ...the lucid & black Rings...ought alwayes to appear after the manner described, were light uniforme. And after that manner, when...two contiguous...glasses have been illustrated in a dark room, by light of any uniforme colour made by a Prism.... but in *compound* light it is otherwise.

---

[74] *Opticks*, Book I, Part 2, Proposition 10. *Optica*, Part II, Lecture 9, *OP* I, 516, 519.
[75] *Opticks*, Book II, Part 2.
[76] *Opticks*, Book II, Part 1, Observation 6.
[77] *Opticks*, Book II, Part 1, Observation 13.
[78] *Hyp., Corr.* I, 383.
[79] *Opticks*, Book II, Part 1, Observation 15.
[80] *Hyp., Corr.* I, 363.
[81] *Hyp., Corr.* I, 370–371.

For the rayes, wch exhibit red & yellow, exciting, as I said, larger pulses in the aether then those which make blew & violet, & consequently makeing bigger circles in a certaine proportion, as I have manifestly found they do, ... the circles made by illustrating the Glasses with white Light, ought not to appeare black & white by turnes ...; but the colours wch compound the white light must display themselves by being reflected, the blew & violet nearer to the center then the red & yellow, whereby every Lucid circle must become violet in the inward verge, red in the outward, & of intermediate colours in the intermediate parts, & be made broader then before, spreading its colours both wayes into those spaces wch I call the black rings ...[82]

Newton even uses the notion of overlapping rings or coloured fringes 'interfering': '...the outward *red* verge of the Second Ring, & inward *violet* one of the third, shall border upon one another (as you may know by computation...) ... & the like edges of the third & fourth rings shall interfere, and those of the fourth & fifth interfere more, & so one.'[83] 'After this the Several Series interfere more and more, and their Colours become more and more intermix'd, till after three or four more revolutions (in which the red and blue predominate by turns) all sorts of Colours are in all places pretty equally blended, and compound an even whiteness.'[84]

Changing the angle of viewing also changes the pattern: 'What has been hitherto said of these Rings, is to be understood of their appearance to an unmoved eye, but if you vary the position of the eye, the more obliquely you look on the glasse, the larger the rings appear.'[85] '...so that the diameter of the same circle is as the secants of the ray's obliquity in the interjected filme of air, or reciprocally as the sines of its obliquity; that is, reciprocally as that part of the motion of that ray in the said film of air which is perpendicular to it, or reciprocally as the force it strikes the refracting surface withal.'[86] 'And as the breadth of every Ring is thus augmented, the dark Intervals must be diminish'd, until the neighbouring Rings become continuous...'[87]

The changing of viewing angle also explained the colour changes observed in the feathers of birds such as peacocks, the colours arising 'from the thinness of the transparent parts of the Feathers; that is, from the slenderness of the very fine Hairs, or *Capillamenta*, which grow out of the sides of the

[82] *Hyp., Corr.* I, 380.
[83] *Hyp., Corr.* I, 381.
[84] *Opticks*, Book II, Part 2.
[85] *Hyp., Corr.* I, 382.
[86] *Of ye Coloured Circles*; Westfall (1965, 191).
[87] *Opticks*, Book II, Part 2.

grosser lateral Branches or Fibres of those Feathers.'[88] Already identified as a thin film effect in Hooke's *Micrographia*, this example of structural colouration, could now be explained with a mathematical theory.

The same variation in pattern results from changing the medium in the film:

> But farther, since...the thickness of Air was to the thickness of Water, which between the same Glasses exhibited the same Colour, as 4 to 3, and...the Colours of thin Bodies are not varied by varying the ambient Medium; the thickness of a Bubble of Water, exhibiting any Colour will be 3/4 of the thickness of Air producing the same Colour. And so according to the same...Observations, the thickness of a Plate of Glass...may be 20/31 of the thickness of Air producing the same Colours; and the like of other Mediums.[89]

'Perhaps it may be a general Rule, that if any other Medium more or less dense than Water be compress'd between the Glasses, their Intervals at the Rings caused thereby will be to their Intervals caused by the interjacent Air, as the Sines are which measure the Refraction made out of that Medium into Air.'[90]

Overall he had no doubt that the phenomenon was of great significance in the explanation of colour:

> The bignesse of the circles made by every colour, & at all obliquities of the eye to the glasses, & the thickness of the Air or intervalls of the glasses, where each circle is made, you will find exprest in...other papers...: where also I have more at large described, how much these Rings interfere or spread into one another; what colours appear in every ring; where they are most lively, where & how much diluted by mixing with the colours of other Rings; and how the contrary colours appear on the back side of the Glasses by the transmitted light, the glasses transmitting light of one colour at the same place where they reflect that of another.[91]

By the time of the *Opticks*, both the periodicity and the disposition to be reflected or refracted, could be seen as an innate and unchangeable property of the light rays: 'what is said of their Refrangibility may be understood also of their Reflexibility, that is, ... their Dispositions to be reflected, some at

---

[88] *Opticks*, Book II, Part 3, Proposition 5.
[89] *Opticks*, Book II, Part 2.
[90] *Opticks*, Book II, Part 1, Observation 10.
[91] *Hyp., Corr.* I, 382.

a greater, and others at a less thickness of thin Plates or Bubbles... are... connate with the Rays, and immutable...'[92]

> ...it appears, that one and the same sort of Rays at equal Angles of Incidence on any thin transparent Plate, is alternately reflected and transmitted for many Successions accordingly as the thickness of the Plate increases in arithmetical Progression of the Numbers, 0, 1, 2, 3, 4, 5, 6, 7, 8, &c...And this alternate Reflexion and Transmission...seems to be propagated from every refracting Surface to all distances without end or limitation... And this action, or disposition, in its propagation, intermits and returns by equal Intervals...[93]

Concerning such phenomena, Newton's general procedure was to adopt a minimalist position, making the fewest possible assumptions compatible with the mathematical description of the phenomenon under investigation. In a typical statement of this position, he writes: 'But whether this Hypothesis be true or false I do not here consider. I content myself with the bare Discovery, that the Rays of Light are by some cause or other alternately disposed to be reflected or refracted for many vicissitudes.'[94] Eventually, this led him to introduce the doctrine of fits of easy reflection and easy refraction, an idea of a purely abstract nature: 'Every Ray of Light in its passage through any refracting Surface is put into a certain transient Constitution or State, which in the progress of the Ray returns at equal Intervals, and disposes the Ray at every return to be easily transmitted through the next refracting Surface, and between the returns to be easily reflected by it.'[95] The fits were certainly not due to some kind of mechanical process: 'the rays are reflected and refracted not by falling upon the solid particles of bodies but by permeating their pores, at short distances from the particles, and...by these actions they create the fits of more easy reflection and more easy transmission.'[96] Rupert Hall thought that 'modulation' might be a more appropriate term than 'fit' to describe their action.[97]

3.4. Wave Theory and Light

Waves or vibrations were important in Newton's theory of light from the beginning, and always remained part of it, and they were always periodic.

---

[92] *Opticks*, Book II, Part 2.

[93] *Opticks*, Book II, Part 3, Proposition 12.

[94] *ibid.*

[95] *ibid.*

[96] *De vi Electrica*, Corr. V, 365.

[97] Hall (1992, 288).

He even used the *term* 'waves' when referring to the coloured fringes in thick or thin films as seen through a prism.[98] What he was never prepared to do was to say that waves actually constituted light, even though that 'hypothesis' would explain many of the phenomena associated with light. In most of his explanations, waves were a kind of secondary phenomenon produced in the medium through which the light travelled by the 'rays' which constituted the light. In his later work, the property of the rays which excited the vibrations became associated with the probabilistic 'fits' of 'easy reflection' and 'easy refraction', which themselves had a periodic property, measured by an 'interval of fits', a kind of wavelength-like property.

The periodicity of the waves was responsible for colour dispersion. Refrangibility and wavelength were closely linked. His classic position on this is stated in Query 13 of the original *Opticks*:

> Do not several sorts of Rays make Vibrations of several bignesses, which according to their bignesses excite Sensations of several Colours, much after the manner that the Vibrations of the Air, according to their several bignesses excite Sensations of several Sounds? And particularly do not the most refrangible Rays excite the shortest Vibrations for making a Sensation of deep violet, the least refrangible the largest for making a Sensation of deep red, and the several intermediate sorts of Rays, Vibrations of several intermediate bignesses to make Sensations of the several intermediate Colours?

Though frequency was unchanged at a refracting boundary, the size of the vibration or wavelength was determined by the refractive index associated with each colour, 'the vibrations excited in ye eye by ye least refrangible rays shall be half as large & twice as many as those excited by the most refrangible.'[99] The wavelength of red light was greater than that for violet. 'The vibrations excited by ye more refrangible rays are shorter then those excited by the less refrangible ones: & upon this difference depends the different reflexibility & different inflexibility of light ...'[100] The wavelengths of light in water and in glass were also different to those in air and related by the refractive index. And: 'in plates of denser transparent bodyes the rings are made at a lesse thicknesse of the plate, (the vibrations I suppose being shorter in rarer æther then in denser).'[101] When periodicity was applied to the theory of fits: 'the Intervals of the fits at equal obliquities are greater

---

[98] *Opticks*, Book II, Part 3, Proposition 13.
[99] Early draft for Book IV, Proposition 10 of *Opticks*, 1690s. Bechler (1973, 21).
[100] Early draft for Book IV of *Opticks*, 1690s. Bechler (1973, 21).
[101] *Hyp., Corr.* I, 383.

and fewer than in the less refrangible Rays, and less and more numerous in the more refrangible ....'[102]

There is a close analogy between the optical waves and those produced by sound, whose wavelengths determine tones or pitch:

> the agitated parts of bodies according to their severall sizes, figure and motions, doe excite Vibrations in the Æther of various depths or bignesses, which being promiscuously propagated through that Medium to our Eyes, effect a Sensation of Light of a white colour; but if by any means those of unequall bignesses be seperated from one another, the largest beget a sensation of a Red colour, the least or shortest of a deep Violet, and the intermediate ones of intermediate colours; much after the manner that bodies according to their severall sizes shapes and motions, excite Vibrations in the Air of various bignesses, which according to those bignesses make severall tones in sound ....[103]

The same analogy applies to the optical waves and those produced in water by a disturbing object like a stone:

> If a Stone be thrown into stagnating Water, the Waves excited thereby continue some time to arise in the place where the Stone fell into the Water, and are propagated from thence in concentrick Circles upon the Surface of the Water to great Distances. And the Vibrations or Tremors excited in the Air by percussion, continue a little time to move from the place of percussion in concentrick Spheres to great distances. And in like manner, when a Ray of Light falls upon the Surface of any pellucid Body, and is there refracted or reflected, may not Waves of Vibrations, or Tremors, be thereby excited in the refracting or reflecting Medium at the point of Incidence, and continue to arise there...? and are not these Vibrations propagated from the point of Incidence to great distances?[104]

He is still arguing this case in the later editions of the *Opticks*, though this time the language has changed to replace 'waves' with 'rays'. Nevertheless, the result is still to produce vibrations:

> Nothing more is requisite for producing all the variety of Colours, and degrees of Refrangibility, than that the Rays of Light be Bodies of different Sizes, the least of which may take violet the weakest and darkest of the Colours, and be more easily diverted by refracting Surfaces from the right Course; and the rest as they are bigger and bigger, may make the stronger and more lucid Colours, blue, green, yellow, and red, and be more and more difficultly diverted. Nothing more is requisite for putting the Rays of

---

[102] *Opticks*, Book II, Part 4.
[103] *Hyp., Corr.* I, 362.
[104] Manuscript connected with *Opticks*, 1690s, Propositions 12–14; Westfall (1971, 412).

Light into Fits of easy Reflexion and easy Transmission, than that they be small Bodies which by their attractive Powers, or some other Force, stir up Vibrations in what they act upon. . .[105]

Newton's demonstration of mathematical periodicity in light waves stands out in stark contrast to the views of contemporaries who are more closely connected with wave theories of light. Hooke, for example, who had been one of the first to observe the effects of periodicity and had devised a pulse theory of light in his *Micrographia*, denied its existence in light waves, though for a brief period in 1675 he seems to have been persuaded by Newton's arguments. On 11 March, he reported to the Royal Society: 'as there are produced in sounds several harmonics by proportionate vibrations, so there are produced in light several curious and pleasant colours, by the proportionate and harmonious motions of vibrations intermingled; those of the one are sensed by the ear, so those of the other are by the eye.'[106] Three years earlier, he had imagined that 'the white or uniforme motion of light' might 'be compounded of all the other colours, as any one strait and uniform motion may be compounded of thousands of compound motions, in the same manner as Descartes explicates the Reason of Refraction,' but he could 'see noe necessity for it.'[107] There is some resemblance here to a theory put forward by L. Georges Gouy 200 years later, which represented white light as the sum of an infinite number of waves of different wavelength or spectral colour, forming single pulses, whose Fourier components could be resolved by a dispersive device such as a prism.[108]

Newton, though normally avoiding the controversies associated with hypotheses, was encouraged by Hooke's 'conversion' into producing an extensive account of his own views on both optics and almost everything else in *An Hypothesis explaining the Properties of Light*. In sending this document to Oldenburg, he wrote:

> I was glad to understand, as I apprehended, from Mr Hooks discourse at my last being at one of your Assemblies, that he had changed his former notion of all colours being compounded of only two Originall ones, made by the two sides of an oblique pulse, & accommodated his Hypothesis to this my sugestion of colours, like sounds, being various, according to the various bignesse of the Pulses.[109]

---

[105] Q 29.
[106] Birch (1756–1757, 3: 194).
[107] Hooke to Oldenburg, 15 February 1672, *Corr.* I, 114.
[108] Gouy (1886).
[109] *Hyp., Corr.* I, 363.

Huygens, who produced a celebrated construction for light waves, based on the production of secondary wavelets at each wavefront, also explicitly denied periodicity when he produced his own version of a theory in which light was transmitted in random longitudinal pulses. 'But as the percussions at the centres of these waves possess no regular succession, it must not be supposed that the waves follow one another at equal distance.'[110] Huygens' wave theory was very different from the one now accepted, and could not have explained such effects as interference, diffraction or polarisation, even in qualitative terms. His waves, unlike Newton's, were not periodic, neither were they transverse; they had no wavelength, frequency, or phase velocity. In his view, for example, the waves or secondary wavelets crossing over each other should not be allowed to interfere. He also specifically denied diffraction, refusing to accept the validity of Grimaldi's experiments, which Newton tried to interpret and extend.

Although the nature of Huygens' innovation has sometimes been construed as lying in the *physical* idea of secondary wavelets, it is, in fact, in a much more subtle mathematical development of which this is a merely crude approximation. The notion of spherical waves acting as a source at each centre of disturbance may have occurred to several authors in the 17th century. Grimaldi, for example, anticipated the kind of argument that he expected his opponents to put forward, by writing: 'you will say that this radiation of *lumen* is such because the air which is illuminated having become like a new source of light creates its own sphere of activity and as a result a secondary *lumen* is produced in multiple ways and in such a manner as to produce these numerous series.'[111] The implication here is that the idea is already well known; and some such idea had previously appeared in 1648 in the *Thaumantias* of Marci of Cronland.[112] It may even be mediaeval in origin.

While making an explicit application of the idea to a wave model, Huygens' significant contribution was a mathematical construction, which, in effect, subverted the physical idea entirely, leading to his own denial of any suggestion of diffraction, interference, or periodicity. Huygens' principle, in truth, was not really concerned with 'physical' wavelets at all. He made the assumption, to justify rectilinear propagation, that the secondary waves would only be perceived at their common tangent. Taken to its logical conclusion, it would make points not on wavefronts assume an intensity

---

[110] Huygens (1912, 16).

[111] Grimaldi (1665, Proposition 1, number 21); Ronchi (1839/1970, 130).

[112] Marci (1648). The mediaeval idea of 'multiplication of species' may be a progenitor.

infinitely greater than those on the wavefronts. Huygens' principle is most certainly not a 'natural' result of wave theory, as is often assumed, but a highly original result of profound mathematical significance, despite being almost a physical impossibility. It can only be made to work for physical waves by making an additional assumption that has a mathematical, but no real physical justification.

3.5. Wave Theory and the Aether

Wave motion suggested an aether:

> ...it is to be supposed, that the Æther is a vibrating Medium like Air; onely the vibrations far more swift & Minute; those of Aire, made by a mans ordinary voice succeeding one another at more than halfe a foot or a foot distance, but those of æther at less distance then the hundred thousand part of an inch. And, as in Air the Vibrations are some larger then others, but yet all equally Swift (for in a ring of Bells the Sound of every tone is heard at two or three miles distance, in the Same Order that the bells are Stroke;) So I suppose the æthereall Vibrations differ in bignesse, but not in Swiftnesse.[113]

The properties required by such an aether were described in a letter to Boyle:

> And first I suppose that there is diffused through all places an æthereal substance capable of contraction & dilatation, strongly elastick, & in a word much like air in all respects, but far more subtile.
> 2 I suppose this æther pervades all gross bodies, but yet so as to stand rarer in their pores then in free spaces, & so much ye rarer as their pores are less.[114]

Classical aethers were, of course, by no means necessarily static. In *An Hypothesis of Light*, Newton says that: '...the æther in all dense bodyes is agitated by continual vibrations & these vibrations cannot be performed without forceing the parts of æther forward & backward from one pore to another by a kind of tremor...'[115] Johann Bernoulli II subsequently developed a theory of an aether, in 1736, in which a multitude of tiny vortices were alternately compressed and rarefied by longitudinal vibrations sent out by sources of light.[116]

---

[113] *Hyp., Corr.* I, 366.
[114] Letter to Boyle, 28 February 1679, *Corr.* II, 289.
[115] *Hyp., Corr.* I, 375.
[116] Bernoulli (1752); Whittaker (1951–1953, 1: 95–96).

Aether offered a vibrating medium to explain the periodic phenomena in thin films. 'For if the Rays endeavour to recede from the densest part of the Vibration, they may be alternately accelerated and retarded by the Vibrations overtaking them.'[117] In his early *Hypothesis of Light*, Newton also put forward a model:

> And so supposeing that Light impingeing on a refracting or reflecting aethe-real Superficies puts it into a vibrating motion; that Physical Superficies being by the perpetual appuls of rays alwayes kept in a vibrating motion, & the aether therein continually expanded & comprest by turnes; if a ray of Light impinge upon it while it is much comprest, I suppose it is then too dense & stiff to let the ray passe through, and so reflects it, but the rayes that impinge on it at other times when it is either expanded by the interval of two vibrations, or not too much comprest & condensed, go through & are refracted.[118]

Newton's early version of an aether was responsible for heating the particles of bodies by continually shaking them:

> Now these Vibrations, besides their use in reflexion & refraction, may be Supposed the cheif meanes, by wch the parts of fermenting or putrifieing Substances, fluid Liquors, or melted burning or other hott bodyes continue in motion, are shaken asunder like a Ship by waves, & dissipated into vapours, exhalations, or Smoake, & Light loosed or excited in those bodyes, & consequently by wch a Body becomes a burning coale, & Smoake, flame, & I suppose, flame is nothing but the particles of Smoake turned by the access of Light & heat to burning Coles little & innumerable.[119]

The *Hypothesis of Light* connected the heating effect with motions generated in the grosser parts of bodies by the mediation of the aether penetrating their *inner* parts:

> For its plaine by the heat wch light produces in bodies, that it is able to put their parts in motion, & much more to heat & put in motion the more tender æther; & its more probable, that it communicates motion to the gross parts of bodies by the mediation of æther then immediately; as, for instance, in the inward parts of Quicksilver, Tin, Silver, & other very Opake bodyes, be generating vibrations that run through them, then by Striking the Outward parts onely without entring the body.[120]

---

[117] Q 17.

[118] *Hyp., Corr.* I, 374.

[119] *Hyp., Corr.* I, 366. A modern vacuum process, the Unruh (or Davies-Unruh) effect, has an acceleration with respect to vacuum leading to a temperature differential.

[120] *Hyp., Corr.* I, 374.

A single ray could generate vibrations all over the body and shake the parts by a resonance effect:

> The Shock of every Single ray may generate many thousand vibrations, & by sending them all over the body, move all the parts, and that perhaps with more motion then it could move one Single part by an Imediate Stroke: for the vibrations by Shaking each particle backward & forward may every time increase its motion, as a Ringer does a bells by often pulling it, & so at length move the particles to a very great degree of agitation wch neither the Simple Shock of a ray nor any other motion in the æther, besides a vibrating one, could do. Thus in Air shut up in a vessell, the motion of its parts causd by heat, how violent soever, is unable to move the bodyes hung in it, with either a trembling or progressive motion; but if Air be put into a vibrating motion by beating a drum or two, it shakes Glass windowes, the whole body of a man & other massy things, especially those of a congruous tone: Yea I have observed it manifestly shake under my feet a cellar'd free stone floor of a large hall, so as I beleive the immediate Stroke of five hundred Drum Sticks could not have done, unless phaps quickly succeeding one another at equal intervals of time. Æthereal vibrations are therefore the best means by wch such a Subtile Agent as Light can Shake the gross particles of Solid bodyes to heat them.[121]

3.6. Double Refraction

Double refraction was the phenomenon, which Huygens succeeded in explaining with a mathematical theory, based on an ingenious hypothesis, though he was unable to extend it to the related property of polarisation. Newton was unable to find a satisfactory explanation or a mathematical result to compete with this, though he did offer a tentative explanation of polarisation, based on analytical reasoning. Double refraction was the phenomenon which ultimately led to the revival of Huygens' wave theory in the early 19th century.

The effect was first reported in 1669 by Bartholin of Copenhagen. In looking through a crystal of Iceland spar (or calcite), brought by a sailor from the Bay of Röerford in Iceland, Bartholin observed that it produced double images of small objects. A ray incident on a crystal of the mineral was refracted twice. One of the refracted rays (the ordinary ray) followed the usual sine law of refraction, first reported by Descartes; the other (the extraordinary ray) followed a law which Bartholin was unable to deduce. In addition, when Bartholin rotated the crystal, he found that the ordinary ray remained stationary, while the extraordinary ray rotated about a normal

---

[121] *Hyp., Corr.* I, 374.

to the crystal facet upon which the light was incident.[122] The discovery seems to have stimulated Huygens to take up the study of light. Huygens certainly knew of the discovery by July 1672 when he began a study of the wave theories by such authors as Hooke and Ignaz Pardies. He corresponded directly with Pardies, receiving manuscripts from him. In the process of this work, he invented his mathematical construction.

Huygens, at first, used his construction to give simple explanations of the processes of reflection and refraction; in the case of refraction, he assumed that light would travel more slowly in a denser medium, and so the whole wavefront representing the common tangent of secondary wavelets at the same phase would be angled in such a way that the light rays would bend towards the normal in the denser medium, as experiment showed. Then, on 6 August 1677, he made the dramatic discovery of his explanation of double refraction, accounting for it mathematically, by assuming that the crystalline structure of the spar produced two kinds of aethereal pulses. While the ordinary ray produced a spheroidal pulse as in the usual form of refraction, the pulse associated with the extraordinary ray was ellipsoidal.

Newton first came into contact with the problem in June 1689, when Huygens, who was particularly interested in meeting the author of the *Principia*, came to England to visit his brother Constantyn. On 12 June, he spoke at the Royal Society about the *Treatise of Light* and the *Discourse on the Cause of Gravity* that he was about to publish. According to the *Royal Society Journal Book*:

> Mr Hugens of Zulichen being present gave an account that he himself was now publishing a Treatise concerning the Cause of Gravity, and another about Refraction giving amongst other things the reasons of the double refracting Island Chrystall.... Mr Newton considering a piece of the Island Chrystall did observe that of the two species wherewith things do appear through that body, the one suffered no refraction, when the visuall Ray came parallel to the oblique sides of the parallelepiped; the other as is usuall in all other transparent bodies suffered none, when the beam came perpendicular to the planes through which the Object appeared.[123]

The first observation had been made by Bartholin, but Huygens had found that it was not entirely true. Two further meetings followed, where

---

[122] Bartholin (1669). Newton includes a mysterious 'crystal' in a table of refractive indices drawn up for his 'Of Refractions' (CUL Add 4000, f. 33ᵛ), which predated Bartholin's publication by a number of years. Though it has the same high refractive index (5/3) as Iceland spar, Shapiro believes it to be an 'imaginary substance' introduced for comparison (*OP* I, 256–257).

[123] *Royal Society Journal Book*, 12 June 1689. Shapiro (1989, 223).

optics was certainly discussed, and Huygens later told Leibniz on 24 August 1690 that Newton had informed him of 'some very beautiful experiments,' probably those on thin films.[124]

Leibniz was greatly impressed by Huygens' work and, during the period when they were still (ostensibly) friendly, tried to stimulate Newton into commenting upon it. Writing to Newton on 7 March 1693, he said: 'I would like your opinion about Huygens's explanation, assuredly a most brilliant one since the law of sines works out so happily.'[125] Newton replied on 13 October without mentioning Huygens' work. He was already involved in his own investigation of refraction.[126] On 26 April 1694, Leibniz wrote to Huygens:

> I have learned from Mr. Fatio, by one of his friends, that Mr. Newton and he have been more than ever led to believe that light consists of bodies which come actually to us from the sun, and that it is in this way that they explain the different refrangibility of light and colours, as if they were primitive bodies which always kept their colours, and which come materially from the sun to us. The thing is not impossible, but it appears to me difficult to understand how by means of these little arrows which, according to them, the sun darts, we can explain the laws of refraction.[127]

It is clear, however, that Newton was troubled with double refraction, and he had a great deal of difficulty in deriving a solution. This meeting with Huygens coincided with a period when Newton was beginning to question the validity of the Cartesian condition for refraction, in which light speeded up in entering the denser medium in proportion to the refractive index, or ratios of the sines of the angles of incidence and refraction. In addition to the difficulties which it faced with respect to normal refraction, the Cartesian condition also seemingly failed to account for double refraction; for it assumed that only perpendicular refractive forces existed, whereas the extraordinary ray produced in double refraction was refracted even when it was perpendicularly incident on a surface, and this would only be possible if the refractive force also incorporated a parallel component. Of course, the sine law of refraction failed in this case, anyway, creating a significant problem for incorporating the phenomenon into a general theory of refraction.

---

[124] *Oeuvres Complètes*, 22 vols., 1888–1950, 9: 471. Shapiro (1989, 224).

[125] *Corr.* III, 248.

[126] Newton to Leibniz, 13 October 1693; *Correspondence*, III, 285–286, translation, 286–287.

[127] Brewster (1855, I, 149–150).

Newton didn't venture into a public discussion of the phenomenon until the Latin *Opticks* of 1706, when in the Query that became Query 25 of the next English edition, he asked:

> Are there not other original Properties of the Rays of Light, besides those already described? An instance of another original Property we have in the refraction of Island crystal... a pellucid, fissile Stone, clear as Water or Crystal of the Rock, and without Colour... if any beam of Light falls either perpendicularly, or in any oblique Angle upon any Surface of this Crystal, it becomes divided into two beams by means of ... double Refraction.[128]

He argued that the result had to come from 'an original difference in the Rays of Light, by means of which some Rays are in this Experiment constantly refracted after the usual manner, and others constantly after the unusual manner.' If the difference was not an 'original Property', 'it would be alter'd by new Modifications in the three following Refractions; whereas it suffers no alteration, but is constant, and has the same effect upon the Rays in all the Refractions.' This was effectively the same principle of conservation he had used in explaining the colour-forming properties of dispersed rays as original to the rays themselves and not due to modifications introduced by the refracting medium. He made his opposition to modification theories clear in Query 27: 'Are not all Hypotheses erroneous which have hitherto been invented for explaining the Phænomena of Light, by new Modifications of the Rays? For those Phænomena depend not upon new Modifications, as has been supposed, but upon the original and unchangeable Properties of the Rays.' And in a draft of Query 21 (later Query 29) he emphasised how the property reinforced the material nature of light: 'its difficult to conceive how the rays of light unless they be bodies can have a permanent vertue in two of their sides which is not in their other sides & this without any regard to their position to the space or medium through which they pass.'[129]

Unable himself to come up with a satisfactory mathematical account of the phenomenon, as he had with other areas of optics, Newton tried devising an empirical rule which worked only at small angles of incidence, a classic example of his relative lack of success at hypothetical reasoning. He did, however, express his opposition, in Query 28, to Huygens' arguments for light as a wave motion, partly on the basis that Huygens' theory required a multiplication of aethers: 'how two *Æthers* can be diffused through all Space, one of which acts upon the other, and by consequence is re-acted upon,

---

[128]'Crystal of the Rock' (or quartz) also showed double refraction, but to a lesser degree.
[129]CUL, Add. 3970.3, f. 258ʳ, NP.

without retarding, shattering, dispersing and confounding one anothers Motions, is inconceivable.' The waves, as such, were not the problem, for waves were a component of his own theory, but Huygens' experiments on Iceland spar seemed to make no sense in the context of his wave theory: 'For Pressions or Motions, propagated from a shining Body through an uniform Medium, must be on all sides alike ....'

Newton, who always believed that his own process of analytical abstraction was the correct procedure in natural philosophy, never accepted a mathematical approach which denied a palpable physical truth, which he clearly believed Huygens had done. To have accepted Huygens' theory would have meant denying the periodicity which he himself had demonstrated, and the equally palpable physical effects of diffraction. Of course, Huygens' construction proved to be a remarkable success when it was taken up by the 19th century wave theorists with a different view of how a wave theory should be constituted; it was an astonishing and brilliant piece of mathematical physics in an area where Newton was unable to find a simple equivalent. It was not, however, an inevitable or absolutely essential component of subsequent wave theory; and it was not the only way of deriving Huygens' explanation of double refraction.

3.7. Interference and Diffraction and Light

Both Newton and Huygens had problems with the effects that we now describe as interference and diffraction. Newton came nearer to a solution. In the case of interference, he had given a comprehensive account of the phenomenon for water waves in his discussion of the tides, but he did not apply it to light in his explanation of the Newton's rings phenomenon. Hooke correctly thought that thin film phenomena required the *combined* effect of two rays, one reflected at the front surface of the film and one refracted through and then reflected at the back surface: 'after two refractions and one reflection, there is propagated a kind of fainter Ray ... this confus'd or *duplicated* pulse, whose strongest part precedes, and whose weakest follows, does produce on the *Retina* ... the sensation of a *Yellow*.'[130] The context of this statement, however, was a hopelessly confused theory of colour, and Hooke also lacked knowledge of the mathematical principle involved.

Nevertheless, Hooke's insight later impressed both Thomas Young and David Brewster. Referring to his own discovery of the principle of interference

---

[130] Hooke (1665, 66).

in light beams, Young wrote that Hooke had made comments, 'which might have led me earlier to a similar opinion.' But, in his opinion, Hooke's failure to adopt a valid theory of colour (Newton's) robbed him of the opportunity of making the breakthrough:

> We are informed by Newton that Hooke was afterwards disposed to adopt his 'suggestion' of the nature of colours; and yet it does not appear that Hooke ever applied that improvement to his explanation of these phenomena, or inquired into the necessary consequence of a change of obliquity, upon the original supposition, otherwise he could not but have discovered a striking coincidence with the measures laid down by Newton from experiment.[131]

For Brewster, also, it was a missed opportunity: 'had Hooke adopted Newton's views of the different refrangibility of light, and applied them to his own theory, he would have left his rival behind in this branch of discovery.'[132]

Newton apparently rejected the idea precisely because it was associated with an erroneous theory of colour. He wrote about Hooke's comment in a draft letter to Oldenburg of 1672: 'it is generally beleived that by the refractions of parallel Superficies no colours are Generated by reason that the second superficies destroyeth the effects of the first, & Mr Hook himself in his *Micrographia* consents to this opinion; the first experiment shall be of the contrary.'[133] In his earlier annotations to the *Micrographia*, he had written: 'And these two strokes of the pulse are effected by reflection on either side of the Muscovy Glasse ye weaker stroke being effected by ye further side of the glasse.'[134]

Newton always knew that two surfaces must be involved, with a degree of phase coherence across the thickness of the film. This was clear from the fact that partial reflection occurred at transparent surfaces: 'the Arithmeticall progression of those thicknesses, wch reflect & transmit the rayes alternately, argues that it depends upon the number of vibrations between the two Superficies of the plate whether the ray shall be reflected or transmitted.'[135] He, also, occasionally tried out some more complicated models, to approach closer to a fit with the experimental phenomena. The early *Hypothesis of Light* has a model with perhaps a suggestion of superposition or interference,

---

[131] Young (1802a, 39–40); Young (1807, 2: 627–628).
[132] Brewster (1855, I, 160).
[133] Draft letter to Oldenburg (1672), *Correspondence*, I, 191.
[134] Annotations to the *Micrographia*, *Unp.*, 403.
[135] *Hyp., Corr.* I, 370–371.

never fully explored:

> Yet all the rayes without exception ought not to be thus reflected or transmitted: for sometimes a ray may be overtaken at the second superficies by the vibrations raised by another collaterall or immediately succeeding ray; wch vibration, being as strong or stronger then its owne, may cause it to be reflected or transmitted when its owne vibration alone would do the contrary. And hence some little light will be reflected from ye black rings wch makes them rather black then totally dark; & some transmitted at the lucid Rings, wch makes the black rings, appearing on the other side of the Glasses, not so black as they would otherwise be.[136]

What Newton never fully grasped, as Young did in the early 19th century but no scientist before him, was the need for a full mathematical combination of waves from two surfaces. The result would have been something like the standing waves Newton had understood for the acoustic waves in pipes, with the nodes and antinodes representing the respective dark and light fringes. Consequently, his values for the 'Interval of the Fits' or equivalent that he derived from the experiment on interference rings are approximately half the values now accepted for optical wavelengths. His accounts of thin film phenomena are otherwise extremely accurate, and incomparably superior to those of any of his contemporaries. More than a century later, Young was able to use Newton's data directly in deriving his values for the wavelengths of different colours of light.

The other main problematic feature of light, which we now call diffraction, was the one that Newton called inflection. Newton seems to have regarded inflection as a special case of the mechanism responsible for refraction and reflection; it could be seen, in the same way as they were, as the result of a force acting to bend the light rays. In Query 28, he looked at the effect in connection with the spreading of waves at obstacles. Here, he wished to reject as erroneous all hypotheses 'in which Light is supposed to consist in Pression or Motion, propagated through a fluid Medium.' 'For in all these Hypotheses the Phænomena of Light have been hitherto explain'd by supposing that they arise from new Modifications of the Rays, which is an erroneous Supposition.'

A theory involving longitudinal pressure waves, like those of sound, acting as a periodic disturbance in a medium, could not explain the spreading of light into the 'quiescent medium' beyond the sharp edges of an object. In one manuscript of the 1690s, however, and in one of the published queries,

---

[136] *Hyp., Corr.* I, 378–379.

there is at least a suggestion that the diffracting force at the edges of bodies results in *transverse* vibrations of the rays of light:

> Prop 12 The motion excited in pellucid bodies by the impulses of the rays of light is a vibrating one & the vibrations are propagated every way in concentric circles from the points of incidence through the bodies.
>
> Prop 13 The like vibrations are excited by ye inflected rays of light in their passage by ye sharp edges of dense bodies: & these vibrations being oblique to the rays do agitate them obliquely so as to cause them to bend forwards & backwards wth an undulating motion like that of an Eele.
>
> Prop 14 As the oblique vibrations excited by inflected rays do agitate the rays sideways wth a reciprocal motion so the perpendicular vibrations excited by the refracted rays do agitate the rays directly with a reciprocal motion so as to accelerate & retard them alternatively & thereby cause them to be alternately refracted & reflected by thin plates of transparent substances for making those many rings of colours described in ye Observations of the third book.[137]

Newton proposes here that, though the vibrations responsible for reflection and refraction are longitudinal, those responsible for diffraction are, in some sense, perpendicular to the motion, and the analogy used in this manuscript for diffraction — the sideways undulations of an eel — is the precise one used in the published Query 3. Diffraction or inflection apparently did show that light could bend at sharp edges, back and forth like an eel, sideways with respect to the motion. At this stage, the vibrations were described as being excited by the rays in the medium, but Newton retained the analogy when he made the vibrations intrinsic properties of the rays themselves.[138]

Of course, we should not assume that these perpendicular vibrations are transverse waves in the modern sense, but transverseness itself was clearly not the main barrier to a final explanation. The persistent use of a water wave analogy suggests that Newton's light vibrations could have been transverse as much as longitudinal, but he also gave the same explanation using purely longitudinal sound waves. Interestingly, Hooke, in his sparring with Newton, again came close to the idea of transverse waves, though again in the context of a hopelessly contrived hypothesis which could not be accepted as a whole.

---

[137] MS connected with *Opticks*, 1690s, Propositions 12–14; quoted in Westfall (1971, 412).

[138] The famous bright 'Poisson spot' at the centre of the shadow formed by an opaque disc was observed at least twice during Newton's lifetime (Delisle, 1715; Maraldo, 1723), but its significance would not be realised for another century.

Writing his first letter to Oldenburg on Newton's optical theory, Hooke proposed that

> the motion of Light in an uniform medium, in wch it is generated, is propagated by simple & uniform pulses or waves, which are at Right angles with the line of Direction; but falling obliquely on the Refracting medium, it Receives an other impression or motion, which disturbs the former motion: — somewt like the vibration of a string; and that, which was before a line, now becomes a triangular superficies, in which the pulse is not propagated at Right angles with its line of Direction, but askew....[139]

It is interesting to note that the oblique wavefronts which Hooke required for a colour theory based on modification meant that he had to adopt the Cartesian refraction condition for phase velocity, which is certainly incorrect, as opposed to the Cartesian condition for particle momentum, which emerged from Newton's theory based on unmodified colours.

Adopting a model of transverse vibrations of small amplitude would have made it possible to fulfil Newton's basic conditions for treating inflection as a form of wave diffraction, and also for explaining polarisation, but we have no reason to suppose that they were considered as the solution by either Newton or any of his contemporaries. Newton's primary concern for the particle-like behaviour he had observed in the process of dispersion by a prism made him accept *intrinsic* wavelike behaviour only where compelled to do so by irrefutable experimental evidence, and then with the minimum of hypothetical comment. Of his contemporaries, Hooke, perhaps, had the best opportunity to produce wavelike theories of both interference and diffraction, both of which he had observed in his own experiments. Hooke, however, was seemingly defeated by the ingenuity of his own hypothetical reasoning, vindicating, to a large extent, Newton's view that hypotheses, though sometimes correct, were basically an unproductive way of approaching science. Though we can, in retrospect, select from Hooke's hypotheses the components that coincide with views that we consider valid today, Hooke himself had no apparent means of doing so. Consequently, he was unable to carry through many of his investigations to a conclusion where he could describe a phenomenon with a fully worked out theoretical resolution. Where Newton could not do this, he took the unresolved aspects to the point where they could be taken up by successors with further experimental or theoretical knowledge, leading to a solution that could be obtained swiftly once the missing pieces were found. In this way, he contributed more than any of his contemporaries to solving the puzzling optical effects that we now describe using the terms interference and diffraction.

---

[139] Hooke to Oldenburg, 15 February 1672, *Corr.* I, 112.

Chapter 4

The Velocity of Light

4.1. The Velocity of Light

There is a statement in Newton's *An Hypothesis of Light*, sent to the Royal Society on 7 December 1675, that should appear more remarkable than it actually does. Here, he writes, in connection with the colours produced during reflection, on a hypothesis that aetherial vibrations were faster than light: 'But to make it the more allowable, it's possible light itself may not be so swift as some are apt to think, for notwithstanding any argument that I know yet to the contrary it may be an houre or two, if not more in moveing from the sunn to us.'[1] At an hour, this figure would only be seven times slower than the time now accepted (499 seconds or 8.317 minutes), or within one order of magnitude, but there doesn't seem to be any particular basis for the figure, just the need to give a round number.

By the time this statement was made, the astronomers Flamsteed and Cassini had each made a good estimate of the solar parallax at 10", equivalent to an Earth–Sun distance of 87 million miles, during the opposition of Mars in 1672, so converting the time delay to a value for the speed of light.[2] However, the significant aspect of the prediction (if we can call it that) is not that the speed would be considerably faster than anything else known on Earth but that the speed may not be *particularly fast* in an astronomical context. At the speed postulated, the delay in an observer receiving information about the Sun's position in space would be clearly non-trivial, and it would be even greater for more distant celestial objects. This would have considerable implications, both for observing celestial motions and for explaining them by fundamental mathematical laws, such as

---

[1] *Hyp., Corr.* I, 377–378.
[2] Flamsteed (1672); Cassini (1730). The results of these two astronomers are specifically cited in Newton's letter to Bentley of 25 February 1693, *Corr.* III, 253.

Newton was at that moment trying to devise. There would always necessarily be a discrepancy between theory and observation.

In fact, within only a few months of Newton submitting the *Hypothesis* to the Royal Society, the Danish astronomer, Ole Rømer (Olaus Roemer), who had been working at the Paris Observatory since 1668, had shown that astronomical observations were indeed affected by the finite speed of light. Rømer's discovery that light has a finite speed of propagation is one of the most significant in the history of physics. Rømer had observed that the eclipses of Jupiter's satellites came later than expected when the Earth was furthest from the planet and earlier than expected when it was nearest. Lecturing to the Académie des Sciences in September 1676, he said that he anticipated that the period of Io, the innermost satellite, would be reduced while the Earth was moving towards Jupiter and increased, six months later, as it moved away. Though a single observation would produce no identifiable reduction, there would be a significant delay over 40 periods. He predicted, in particular, that there would be a delay of 10 minutes in the satellite emerging from the shadow of the planet on 9 November, and this prediction turned out to be successful when the eclipse was observed.

In the only publication of his result,[3] Rømer did not give a speed, rather a lower limit in today's units of $140,000 \, \text{kms}^{-1}$. He said that it covered a distance like Earth's diameter, about 3,000 lieues (that is, $3,000 \times 4.448 \, \text{km}$), in less than one second, and that it would take about 11 minutes for light from the Sun to reach the Earth. Cassini, the Observatory's Director, was at first supportive, but when he found himself unable to find the same effect with Jupiter's other three satellites, he became critical. (There is no substance to the claim that Cassini had found the result previously in 1675.) Rømer's paper was translated in the *Philosophical Transactions* in July 1677, and he visited Greenwich in 1679 and 1688.[4] Flamsteed, the Astronomer Royal, was in favour, Hooke, the Royal Society curator, opposed. In 1680, he argued: ''tis so exceeding swift that 'tis beyond Imagination; for so far he thinks indubitable, that it moves a Space equal to the Diameter of the Earth, or near 8,000 miles, in less than one single Second of the time, which is in as short time as one can well pronounce 1, 2, 3, 4: And if so, why not be as well instantaneous I know no reason....'[5]

---

[3] Rømer (1676).
[4] Rømer (1677).
[5] Hooke (1680), *Posthumous Works*, 78; further references in lectures of May 1680 and 1682, *Posthumous Works*, 108, 130. Wróblewski (1985, 625–626).

Rømer also sent his result to Christiaan Huygens, who was at the time also working in Paris. Huygens, like Newton, was predisposed to think that the velocity of light was finite, considering it a proof of the wave theory he was developing — it is certainly a necessary condition. He learned of Rømer's discovery of the true velocity about a month after he made his most significant application of the wave theory, his explanation of double refraction, and he wrote to Rømer on 11 November 1677 describing his pleasure at finding that the astronomer had demonstrated what he had merely assumed, especially in his explanation of the refraction in Iceland spar.[6]

Huygens made the first calculation of an actual speed of $16\frac{2}{3}$ earthdiameters per second (or 22 minutes for the Earth's orbit), in his *Traité de la Lumière*, presented to the Académie in 1678, but not published until 1690.[7] Using the value of 12,750 km for the Earth's diameter and estimating the diameter of its orbit as 24,000 Earth diameters, Huygens's data suggested $212,400 \, \text{kms}^{-1}$, about 2/3 of the actual value. Without the rounding, his speed of light would have been $232,000 \, \text{kms}^{-1}$. Huygens's calculation was made using an *emission* model; the calculation within the wave model would require the knowledge of the speed of both the Earth and Jupiter relative to the aether, information which the experiment of Michelson and Morley later found to be impossible to acquire from an optical experiment.

A letter from Flamsteed to Newton, dated 1684, mentioning 'Roemers equation of light,' provides the first connection of Rømer's discovery with Newton.[8] Newton, like Huygens, became an early supporter of a result which many at the time considered dubious. In the first edition of the *Principia*, he used the 22 minute figure provided by Rømer.[9] Halley, however, in 1694, found that Rømer's 22 minutes should have been 17 minutes, leading to a light speed of approximately $300,000 \, \text{kms}^{-1}$.[10] Newton's first good value is found in the *Opticks* of 1704, where he says that it takes only 8 minutes for the light to travel from the Sun to the Earth: 'Sounds move about 1140 *English* Feet in a second Minute of Time, and in seven or eight Minutes of Time they move about 100 *English* Miles. Light moves from the Sun to us in about seven or eight Minutes of Time, which distance is about 70,000,000 *English* Miles, supposing the horizontal Parallax of the Sun to

---

[6]Huygens (1677). Shapiro (1974, 218).

[7]Huygens (1912, 7–10); Wróblewski (1985, 625).

[8]Flamsteed to Newton, 27 December 1684, *Corr.* III, 403–405.

[9]*Principia*, I, Proposition 96, Scholium, first edition, 1687. Here, Newton says light takes about 10 minutes to travel from Sun to Earth.

[10]Halley (1694). Halley's time for the semidiameter was 8.5 minutes.

be about 12".' 'This was observed first by *Roemer*, and then by others, by means of the Eclipses of the Satellites of *Jupiter*.'[11] Newton's estimate for c is around $241,000 \, \mathrm{kms^{-1}}$. The Scholium to Proposition 96 of the second edition of the *Principia* includes the statement 'For it is now certain from the phenomenon of *Jupiter's* Satellites, confirmed by the observations of different astronomers, that light is propagated in succession, and requires about seven or eight minutes to travel from the sun to the earth.'

Rømer's brilliant inference thus resolved a question which had been debated since antiquity, over whether light was transmitted instantaneously or in a finite time. Opinion on the question among earlier writers had been fairly evenly divided. Empedocles of Agrigento, Ibn al-Haytham and Roger Bacon, for instance, believed in a finite light speed[12]; Descartes (though his statements are ambiguous) and Kepler believed that light transmission was instantaneous.[13] Galileo memorably failed to get a measurable result — that is, if he actually tried the experiment with lanterns on mountain tops that he wrote about.[14] Both emission and wave theories would seem to require a finite speed, as would any explanation of refraction that involved a change in velocity at a boundary between two media. So, it is not surprising that Rømer's result was enthusiastically welcomed by both Newton and Huygens. Both men considered the finite velocity of light essential for their very different optical theories. Huygens seemingly felt that it provided support for his wave theory, though he used an emission theory to calculate his numerical value. Newton *had* to assume that light had a finite speed for his theory of light vibrations to be true, as well as for the geometrical analysis of dispersion in his optical lectures, though his own quoted figure for the speed from 1675 seems to be only an offhand comment, without any real authority.

For about half a century, however, Rømer's result remained the only indication that light had a finite speed. All of Newton's calculations involving c relied on measurements of the orbits of Jupiter's satellites. But, at the end of Newton's life, the astronomer James Bradley found another astronomical effect which involved the finite velocity of light and a discrepancy between

---

[11] Book II, Part 3, Proposition 11.

[12] Ronchi (1939/1970, 3–8, 13–15), discusses the early theories of light in detail. Sarton (1993, 248) (for Empedocles). Alhazen (1572, IV, 17–18, 112–113); Sabra (1967, 72–78); Lindberg (1974, 396; 1996, 143).

[13] Descartes (1634); MacKay and Oldford (2000); Wróblewski (1985, 624); Descartes (1637, 6, 84); Sabra (1967, 55). Descartes argued that a finite speed of light would refute his whole theory. Sakellariadis (1982, 1); Kepler (1604).

[14] Galileo (1638/1954, 42–44); Boyer (1941); Wróblewski (1985, 624).

the positioning of celestial bodies and their observation. This time it involved distant stars and allowed a much more accurate measurement of light speed. Essentially, Bradley found a shift in the positions of stars from what he expected, which was not due to the long-sought after stellar parallax, but which, according to the emission theory, was due to the time it took for light to travel down the telescope tube. The first indication of the effect, which Bradley called 'aberration', had been observed by December 1726 and was noted as a newly-discovered anomaly in Newton's late correspondence, but the explanation was only published in 1728, a year after Newton's death.[15] Interestingly, Hooke had seen the effect in 1669 in the star γ Draconis but had mistakenly attributed it to stellar paradox.[16] Bradley's discovery, in addition to pointing to a necessary correction in the measurement positions, established the validity of Rømer's explanation of his observations of the satellites of Jupiter, and provided the first precision value for c. It also indicated that the velocity of light was a constant, that is, not dependent on the velocity of the source, as the aberration angle v/c, measured with respect to the velocity of the Earth v, was observed to be constant within the limits of observation. It would be more than a century before the velocity of light could be measured in a terrestrial experiment.

4.2. The Velocity of Light and Refraction

Even before the experimental discovery of its finite value, it had become apparent that the velocity of light provided a way of understanding the sine law of refraction, first published by Descartes but also known as Snell's Law, that the ratio of the sines of the angles of incidence and refraction at the boundary of two media $(\sin i/\sin r)$ was a constant for the media, now known as the refractive index (n).[17] This was because the sine law could be explained as indicating a *change* of speed as light crossed the boundary from one medium to the other. However, it was by no means obvious whether it increased or decreased in speed as it passed from a less dense medium (such as air) to a denser medium (such as water). A special result of the sine law was that light travelling from a denser to a less dense medium and striking the boundary at an angle greater than a certain critical value, or *critical angle*, would undergo *total internal reflection* within the denser medium and not succeed in entering the less dense medium at all.

---

[15]Bradley (1728).
[16]Hooke (1674).
[17]Descartes (1637).

On this Newton writes:

> Thus for instance it may be observed in a triangular glass Prism..., that
> the rays... that tend out of the glass into Air, do by inclining them more &
> more to the refracting Superficies, emerge more & more obliquely till they
> be infinitely oblique, that is in a manner parallel to the Superficies, wch
> happens when the angle of incidence is about 40 degrees, and that if they
> be a little more inclined are all reflected... becoming I suppose parallel to
> the Superficies before they can gett through it.[18]

'...in the... Experiment, by turning the Prism about its Axis, until
the Rays within it... became... totally reflected...; those Rays became
first of all totally reflected, which before at equal Incidences with the
rest had suffered the greatest Refraction.'[19] The experiment provided a
clearer demonstration than refraction from a prism that light rays producing
different colours had immutable properties and could be separated out of
the heterogeneous mixture composing white light, with rays of the greatest
refrangibility also having the greatest reflexibility.

The fact that the critical angle for a ray depended uniquely on the
colour associated with it not only provided a more clear-cut demonstration
of the immutability of colour than a further refraction, but also produced
a 'striking' visual effect, which may perhaps be classified with all those
special optical effects associated with the phenomena of reflection, refraction,
diffraction and polarisation, that are variously designated by such names
as fringes, brushes, bands, or ghosts. This was a blue-coloured bow which
appeared at the base of a prism as a result of the rays constituting daylight
being selectively reflected there.[20]

The two possible outcomes for the velocity ratio led to two classes of
optical theory, each which gave a complete mathematical explanation of the
law of refraction. Contrary to many statements in the literature, they did not
divide simply into particle and wave theories, though that is how they were
perceived as attitudes hardened in later centuries. One class of theory was
initiated by Descartes himself, though it is difficult to relate it to an actual
ratio of velocities. Descartes seems to have thought that the propagation
of light was instantaneous and that v_1 and v_2, which we would describe as

---

[18] *Hyp., Corr.* I, 372.

[19] *Opticks*, 1704, Book I, Part 1, Proposition 3, Theorem 3. The experiment was first
described in *Of Colours*, §22, CUL Add 3975, f. 4ᵛ. Newton measured the critical angle
from air to glass at $40°16'$ (*Optica*, II, Lecture 12, 116, *OP* I, 560–561), allowing him to
use 45° prisms as total reflectors.

[20] See Shapiro, *OP* I, 35. *Of Colours*; *Optica* II, Lecture 12, 116, *OP* I, 558–561.

the respective velocities in media 1 and 2, merely represented *tendencies* to motion. From a notebook entry of 1619–1621,[21] which may or may not have preceded his discovery of the refraction law, we know that Descartes was the first to assert that the passage of light would be easier through a denser medium, in contradiction to all earlier authorities, leading to the refraction condition:

$$\frac{\sin i}{\sin r} = \frac{v_r}{v_i}.$$

In the commonly observed cases, such as refraction from air to water or air to glass, this would mean that light travelled faster through water or glass (the denser medium) than it did through air. The modern way to explain this is to say that light experiences a force normal to the refracting surface, leading to its velocity component being accelerated in that direction, the component parallel to the surface remaining unchanged. Since light bends towards the normal in the denser medium, then the velocity must be greater in that medium, and the refraction condition can be found from the horizontal components where $v_i \sin i = v_r \sin r$. The argument became associated with supporters of a particulate theory of light, and also with Newton, who expressed it in this form in Proposition 94 of the first book of the *Principia*, the only time it appeared in his published work.[22] However, it was neither specific to particle theorists, which would include Descartes only in a rather special sense,[23] nor a required view among them.

The alternative condition, which has been mainly associated with wave theorists (but again neither specific to nor required among this group) was that the speed of light would be reduced in the denser medium, leading to the refraction condition

$$\frac{\sin i}{\sin r} = \frac{v_i}{v_r}.$$

The first supporter of this viewpoint was the philosopher-physicist, Thomas Hobbes, who was evidently a more distinguished mathematician than his reputation as a 'circle-squarer' would suggest. It requires, in the modern interpretation, the idea that rays of light are normal to what are now called

---

[21] Descartes (1619–1621).

[22] In the unpublished *Lectiones Opticae*, he included a version of Descartes' derivation for the plane of refraction, extending it 'to any plane perpendicular to the refracting surface' (Lecture 1, 6, Shapiro, *OP* I, 56–57).

[23] Descartes' theory pictured light as a continuous pressure pulse creating rotational motion in minute globules of a material aether, so the 'particle' aspect referred to the medium rather than to light.

wavefronts.[24] As the angle the wavefront makes with the normal decreases in the denser medium, so does the distance travelled by the wavefront in the same period of time, and therefore the speed of the light waves. For the case of a passage from air to water, where the refractive index is 4/3, Descartes' refraction condition would make light travel 4/3 times faster in water than in air, whereas Hobbes's would make it travel 3/4 times slower.

It had long been established that there were only two real options for a theory of the propagation of light, discrete or continuous, subsequently, more specifically, particles and waves, and 17th-century theorists divided more or less equally between them. Hobbes became the first of the kinematic wave or pulse theorists, making a clear distinction between the motion of bodies, or material particles, and the motion of pulses. He had moved from being a particle theorist himself to reject this position in his *Tractatus Opticus* of 1641. However, it was not necessary to be a wave theorist to opt for his refraction condition. The particle or 'emission' theorists Maignan, in 1648,[25] and Barrow, in 1668,[26] followed Hobbes, while improving significantly on his sine law derivation. They used the Hobbes refraction condition rather than Descartes,' because the corpuscular elements in their theories were counteracted by their assumptions of fluidity in the collective behaviour of their light particles. Barrow's theory appeared in *Lectiones Opticae* (1669), a work to which Newton made some small mathematical contributions.

The wave theorists who followed, including Grimaldi, Hooke, Pardies, Ango and Huygens, read either Maignan or Barrow, and were influenced by them. Hooke, however, replaced Hobbes's assumption of alternate expansion and contraction of the entire source with independent vibrations of the individual particles, and, to accommodate his theory of colour modification, he opted for the alternative refraction condition by drawing his refracted wavefronts at an angle to the rays.[27] Pardies and Huygens were the first to define rays as mere geometrical lines; Newton conceived them as physical objects. For him, the rays were both particles and mathematical constructs, though the mathematical aspects of his optics did not require making physical assumptions. Pardies' rays were drawn normal to spherical wavefronts, but Huygens' differed in being drawn from the original centre of the wave motion to the points of each of the secondary wavelets on their

---

[24]Hobbes (1644, 576–587). Shapiro (1974, 143–155, 172–181, 258–263).
[25]Maignan (1648).
[26]Barrow (1669).
[27]Grimaldi (1665); Hooke (1665). Pardies' lost work is reported in Ango (1682). Huygens (1912, 35–39); Shapiro (1974, 143–155, 172–181, 258–263).

common tangents. The significance of this innovation only emerged when Huygens introduced wavefronts that were non-spherical.

In modern versions of the wave theory, as in the Hobbes–Huygens tradition, we understand that refraction is due to the different speeds of light waves in different media. Light waves travel more slowly through optically denser media and are consequently refracted more. The refraction is measured by defining the refractive index at the boundary between two media as the ratio of the sines of the angles of incidence and refraction, $n_{ir} = \sin i / \sin r$. This ratio is also the same as the velocities of waves of light in the respective media, v_i / v_r or v_1 / v_2. The more refracted waves are also the slower ones. The absolute refractive index of a medium, which measures the amount of slowing down in a more refracting medium, is defined as the ratio of the velocity of light in vacuum (c) to the velocity of light waves in the medium (v). Light in the medium actually travels at c but is slowed down by successive absorptions and re-emissions.

A significant addition to the Hobbesian condition was made when Pierre de Fermat showed that the law of refraction could be derived by assuming that the path of light taken was the one that took the least time.[28] Huygens' theory, of 1678, which is now considered as the pinnacle of the kinematic wave approach, showed that the Hobbesian condition followed from his wavefront construction combined with Fermat's principle of least time. Inspired by Rømer's discovery of light's finite speed, Huygens, at first, used his construction to give simple explanations of the processes of reflection and refraction; for refraction, he assumed that light would travel more slowly in a denser medium, and so the whole wavefront representing the common tangent of secondary wavelets at the same phase would be angled in such a way that the light rays would bend towards the normal in the denser medium, as experiment showed. As we have seen, he subsequently used his construction in the explanation of double refraction, where Snell's law no longer applies.

It has often been assumed that Newton's position in this developing scenario was to support the Cartesian condition, and a historical tradition developed, largely created by Young, in which Newton, as a supporter of the particle theory and the Cartesian refraction condition, was set up in opposition to Huygens, a supporter of the wave theory and the Hobbesian condition.[29] Eventually, it was claimed, that the experiments of Foucault, in 1850, in measuring the speeds of light in air and water decided the issue in

---

[28] Fermat (1657, 1662).
[29] Young (1800, 1802b, 1807).

favour of the latter.[30] However, it is now clear that this picture, though polemically convenient for Young, has no historical validity. Apart from falsely representing the *physical* picture, it is also incorrect in a historical sense.

Newton, certainly, is incorrectly represented as supporting any particular hypothesis of light. He may have inclined towards a particle theory, on the basis of his work on dispersion through a prism, but it was, for him, only a hypothesis and not to be regarded on the same level as a mathematical theory, such as universal gravitation. His theory also included wave aspects which Huygens' 'wave' theory lacked, in particular, periodicity, frequency, wavelength, and wave velocity, which were vital to Young, and experimental investigations of interference and diffraction, which were the very features that Young wanted most to explain, but which were necessarily eliminated from Huygens' theory. In Proposition 94 of the first book of the *Principia*, Newton derived a version of the Cartesian condition for refraction, but only as a general mathematical theorem for particles acting in a certain way which may or may not have the same properties as supposed particles of light, and, in the *Opticks*, there is no derivation of this condition at all. This is not a coincidence, as it depended on experimental tests which Newton chose not to make public. These facts make it unwise to choose Newton as a confirmed champion of the Cartesian condition. Interestingly, in 1746, Euler, who translated many of the propositions of the *Principia* from geometrical to algebraic form, succeeded in using Newton's own theory of waves to derive the Hobbesian refraction condition by the modern textbook method.[31]

4.3. The Velocity of Light and Dispersion

As is well known, Newton made a major discovery in optics when he observed the elongated spectrum of colours formed by a prism and attributed this dispersion to the fact that rays of light producing the sensations of different colours in the eye had different refractive indices and therefore emerged from the prism at different angles. He showed that the colour-forming properties were innate to the light rays and not due to the structure of the prism, by separating out rays of different colour and noticing that they were unchanged in passing through a second prism. In another version of this experiment, he produced a purer spectrum by focusing it, using a smaller hole for the solar light source together with a lens placed in front of the first prism, after

---

[30]Foucault (1850, 1862).
[31]Euler (1746).

which the separate, purer, colours passed unchanged through the second prism. He also rotated a wheel constructed so that only a single colour could be observed at any time, and there would be no mixing. Rotating the wheel rapidly still produced a sensation of whiteness in the eye of the observer, which could not be due to the *colours themselves* mixing. Newton thus attributed the sensation of 'whiteness' to a combination of all the spectral colours to which the eye was sensitive, and showed that both a lens and a second prism could be used to reverse the dispersion and reconstitute white light. White light could also be produced by overlapping the spectra from two prisms placed side by side, though removing one of them produced colour again. Taken together, this ordered sequence of experiments and inferences constituted a breakthrough in optical science which, in characteristically Newtonian fashion, was greater than the sum of its parts.

In the *Optical Lecture*, Newton writes of colours being produced by selective reflection: '... I find, that all colours of all bodies are generated from nothing else but a disposition, whereby they are disposed to reflect some rays and to let others pass.'[32] However, in the ninth lecture of *Optica*, Part II,[33] Newton, according to Shapiro, 'introduced into optics the concept of selective absorption (*potestas suffocandi et in se terminandi*) as an independent physical process in the coloration of natural bodies,' when he wrote that 'there is in bodies not only a power to reflect or transmit rays, but also one to stifle and terminate them within themselves.'[34] According to Shapiro: 'Selective absorption — and not, as Newton would have it, selective reflection — is the principal process involved in the color of most natural bodies.' Together with the concept of spectroscopy, it eventually led to absorption spectroscopy, one of the most important tools in both chemical analyses and in astronomy.

Newton continues:

Thus, some bodies stop and retain all kinds of rays, and, in that way they become completely black; others reflect some and suppress the rest, as opaque coloured bodies; others suppress some and partly reflect and partly transmit the rest, as transparent colored bodies that are the same color all around, and others reflect some and transmit the rest, as is established by the examples just related; and so on.[35]

---

[32]Lohne (1965, 137); *Optica*, II, Lecture 1, 5, *OP* I, 439: 'I find that all colors of bodies are produced in no other way but from a certain disposition whereby they are disposed to reflect some rays and to let in others.'

[33] *Optica*, II, Lecture 9, 74, *OP* I, 512–515; Shapiro (1993, 107–108).

[34] *Optica*, II, Lecture 9, 74, *OP* I, Latin, 512, transl. 513.

[35] *Optica*, II, Lecture 9, 74, *OP* I, Latin, 512, 514; transl. 513, 515.

The discovery of the dispersion of white light added another compli-cation, for it implied that the rays producing the sensations of different colours, and therefore with different refractive indices, must also be travelling at different speeds in a dispersive medium. In the *Quaestiones Quaedam Philosophica*, Newton writes: 'Note that slowly moved rays are refracted more then swift ones.'[36] And he extends this to use velocity as an explanation of chromatic dispersion. In the words of Lohne: 'Blue rays are reflected more than red rays, because they are slower. Each colour is caused by uniformly moved globuli. The uniform motion which gives the sensation of one colour is different from the motion which gives the sensation of any other colour.'[37] In Newton's own words: 'slowly moved rays being seperated from ye swift ones by refraction, there arise 2 kinds of colours viz: from ye slow ones blew, sky colour, & purples, from ye swift ones red.' 'Hence rednes yellownes &c are made in bodys by stoping the slowly moved rays without much hindering of the motion of the swifter rays, & blew green & purple by diminishing the motion of the swifter rays, & not of the slower.'[38]

At the time when these statements were made, Newton was attending Barrow's mathematical lectures (the ones on optics date only from 1668– 1669) and it is not impossible that he was using a Maignan–Barrow-type mathematical analysis of refraction to explain his own discovery of dispersion through a variable index of refraction, though there is no evidence of any direct influence from Barrow. Alternatively, he may have been responding to Descartes who had made the red globules in his aethereal medium *rotate* faster than blue when stimulated by the pressure pulse that constituted light. However, since he had not yet read the relevant passages in Descartes, another, and perhaps more probable, explanation is that Newton was responding to his own early experiments on coloured fringes produced between thin plates which he had discovered independently of any suggestion by Hooke, but which he may have interpreted in the light of the pulse theory suggested in Hooke's *Micrographia*. They are certainly written within the Hobbesian refraction condition (supported by Barrow but not yet used by Huygens), and add the additional fact that, within this way of thinking, red light travels faster and is less refracted than blue.

---

[36] QQP, c 1664, 'Of Colors', 122^v, NP.

[37] Lohne (1965, 133).

[38] QQP, c 1664, *Of Attomes*, f. 122^v, *Of Colours*, 122^v, NP. Newton also says: 'The more uniformly the globuli move the optick nerves the more bodys seme to be coloured red yellow blew greene &c but the more variously they move them the more bodys appear white black or Greys' ('Of Attomes', 122^r).

Newton's subsequent response to the question of dispersion and its connection with refraction took a very complex turn, related to a continuing attempt to put optical theory on the same basis as mechanics. This was later largely abandoned, leaving optics more as a descriptive science, whose full mathematical treatment would have to await subsequent discoveries. In this process, he tried out two 'laws' of dispersion (quadratic and linear), neither of which worked (the fate of most hypothetical models in science). It is clear, however, that, for Newton, mathematical techniques, though immensely powerful, also had their limitations, for he was never able to devise a mathematical structure for optics that could stand alongside the one with which he revolutionised mechanics. As he explained in a letter to Oldenburg, mathematical development was valid only when it was founded on correct physical principles, and the exactness of a mathematical demonstration was not a proof of its physical truth: 'I said indeed that the *Science of Colours was Mathematicall & as certain as any other part of Optiques*; but who knows not that Optiques & many other Mathematicall Sciences depend as well on Physicall Principles as on Mathematicall Demonstrations: And the absolute certainty of a Science cannot exceed the certainty of its Principles.'[39] As these comments suggest, Newton, though always privileging mathematical demonstration as the most desirable form of scientific explanation, recognised the danger that mathematical analyses could give an impression of certainty where none really existed, and he, for one, was not deceived by any of his own mathematical models for optics, including those demonstrating the sine law of refraction.

Although he moved from the Hobbesian to the Cartesian condition of refraction, when he tried to explain Snell's law on the basis of forces acting on light corpuscles (c 1670),[40] his position concerning these developments was always equivocal and he came, in a sense, to reject — though never in explicit public terms — even the simple Cartesian condition, when it led to difficulties with a velocity-related interpretation of dispersion. Though a firm believer in the probability of a particulate nature of light, he had no commitment to any particular dynamic model for these processes. He tried many different explanations without finding any that fitted all the observed results. He finally rejected the mathematical derivation of the sine law of refraction via the Cartesian condition, basing his rejection on an experiment which has remarkable similarities to one that seemingly led to a definitive answer more than 160 years later.

---

[39] Newton to Oldenburg, 11 June 1672, *Corr.* I, 187.

[40] *Lectiones Opticae*, Lecture 1, 6, *OP* I, 54–57; *Optica*, I, Lecture 1, 6, *OP*, I, 288–291.

Newton, in 1664–1665, had clearly recognised that a *velocity-distribution* model would explain 'all his Theory' of colours, yet immediately after the first entry on this in his notebook, he replaced it with a second theory, in which the dispersion of light in a refracting medium was the result of a distribution over the spectrum of light-particle *masses*.[41] The reason for the change seems to have been Newton's realisation of the importance of conservation principles. Velocity, unlike mass, was not an immutable or conserved quantity. Optical experiments, however, had shown that the colours (or rather the colour-producing properties) of light rays *remained unchanged or unmodified* even when subjected to processes such as reflection and refraction. This was the very reason why Newton insisted on some kind of particle nature for the rays of light and never accepted a continuous wave theory: 'Colours & refractions depend not on new modifications of light but on the original & unchangeable properties of its rays & such properties are best conserved in bodies projected. Pressions & motions are apt to receive new modifications in passing through several mediums but the properties of bodies projected will scarce be altered thereby.'[42]

His early optical work had already led him to believe that reflection and refraction could not be due to simple mechanical collisions between light particles and molecules within the reflecting or refracting surface, because it was against all probability that all the particles on the surface of a solid, still less that of a liquid, would always be oriented in the same direction. He, therefore, supposed that reflection or refraction, like gravity at the surface of the Earth, was due to a power acting over the surface as a whole. 'The Cause of Reflexion is not the impinging of Light on the solid or impervious parts of Bodies, as is commonly believed,' 'but some other power by which those solid parts act on Light at a distance.'[43] Reflection was 'effected, not by a single point of the reflecting Body, but by some power of the Body which is evenly diffused all over its Surface, and by which it acts upon the Ray without immediate Contact.'

Once he had opted for the mass-distribution model, Newton immediately set about applying the laws he had recently discovered for particle dynamics, that is the conservation of momentum and the law of restitution after impact (equivalent, in the case of perfect elasticity, to the conservation of kinetic energy). He developed a model for dispersion which was incorporated into his optical lectures a few years later, and held to it for the greater part

---

[41]Cambridge University Library, Add. 3996 f. 72^v; Hall (1948).

[42]Draft Q 21/29 for *Opticks*, 1706; Bechler (1973, 33).

[43]*Opticks*, Book II, Part 3, Proposition 8.

of his life, although with subtle alterations.[44] A very late paper, written mainly by Desaguliers, but seemingly with contributions by Newton himself, includes the passage: 'The most refrangible rays consist of smaller particles than the least refrangible rays, & therefore must have least momentum; & consequently are more easily turned out of the way by attraction or repulsion, which makes the curves made by the purple & violet rays … to be less and nearer the said surface than the curves made by red & orange rays … .'[45]

According to John Hendry, the purely geometrical diagrams in the lectures imply, as in this passage, a *momentum* whose component perpendicular to the surface gains or loses a fixed amount but whose parallel component remains unchanged, not, as in the traditional account, a *velocity*.[46] This is a very significant distinction, for, according to the modern quantum theory, the wavelength of light is reduced in the medium by the factor n, the refractive index. Reducing the wavelength λ increases the momentum, which is the ratio of energy E to the phase wave velocity v. Assuming constant frequency ν,

$$p = \frac{h}{\lambda} = \frac{h\nu n}{c} = \frac{h\nu}{v}$$

and

$$E = pv.$$

The result of this is that the argument which gives a Hobbesian refraction condition when taken in terms of the phase velocity, or velocity of an individual wave, gives a Cartesian refraction condition when taken in terms of the particle momentum. The momentum for a light quantum in a refracting medium is, thus, not simply equal to the product of its mass and the wave or signal velocity. In the course of his work, Newton found that there was definitely a discrepancy, though a definitive answer eluded him. It meant that he went through almost every conceivable theoretical position in his attempt to explain this phenomenon. In his earliest work, he proposed that red light was faster than blue; later he claimed that blue moved faster than red, before finally adopting the position that the rays responsible for producing light of different colours must all move at the same speed. It is not always clear, in these theories, whether the velocities are to be measured in dispersive media or in vacuum, and how the velocities involved in these

---

[44]Bechler (1973, 3–5); *Lectiones opticae*, Lecture 11, 115–123, *OP* I, 198–211; *Optica*, I, Lecture 7, 42–50, *OP* I, 334–347; *MP* III, 466–470.
[45]Desaguliers (1728); Bechler (1973, 36).
[46]Hendry (1980, 244–246).

two physical conditions would then be related. The remarkable thing is that, in different ways, all these theoretical positions can be justified.

4.4. The Inversion of Velocities

Newton's early work on interference rings had shown that, for oblique incidence of the light, there was some kind of reciprocal relationship between momentum ('motion') of the incident light corpuscles and the thickness of the film producing rings of each colour (and so, in effect, the associated wavelength): 'What has been hitherto said of these Rings, is to be understood of their appearance to an unmoved eye, but if you vary the position of the eye, the more obliquely you look on the glasse, the larger the rings appear.'[47] 'soe y^t the diameter of ye same circle is as ye secants of ye rays obliquity in ye interjected filme of aire, or reciprocally as ye sines of its obliquity; that is, reciprocally as y^t part of the motion of y^t ray in ye said filme of aire wch is perpendicular to it, or reciprocally as ye force it strikes ye refracting surface wth all.'[48] The same was true when the air in the gap between plate and lens was replaced by a denser medium such as water.

Newton believed that light corpuscles would move faster through water than through air because the aether was rarer in a denser medium; at the same time, his experiments indicated that the associated aether waves had *shorter* wavelengths. Applying his knowledge of wave theory, in particular the constancy of the frequency, Newton realised that this would mean that waves in water actually travelled slower, and proportionately according to the refractive index as the corpuscles moved faster. So the respective velocities of waves and corpuscles, v and u, would be related by the equations

$$\frac{u}{n} = \frac{n}{v}$$

or

$$\frac{u}{v} = n^2$$

or, in modern terms,

$$\frac{u}{c} = \frac{c}{v}.$$

This is the equation linking the direct and inverse (or Cartesian and Hobbesian) refraction conditions, derived in principle from the relation

---

[47] *Hyp., Corr.* I, 382.
[48] *Of ye Coloured Circles*, Westfall (1965, 191).

de Broglie arrived at in 1924:

$$p = \frac{h}{\lambda}.$$

Though Newton never achieved a consistent pattern in his explanation, often showing significantly contradictory statements in the same paper, his early work on the Newton's rings pattern formed in different media led to the hypothesis 'That if y^e medium twixt y^e glasses bee changed y^e bignes of ye circles are also changed. Namely to an eye held perpendicularly over them, the difference of their areas (or ye thicknesses of ye interjected medium belonging to each circle) are reciprocally as ye subtlity of ye interjected medium or as ye motions of ye rays in that medium.'[49] Corpuscles travelling faster in water in relation to the refractive index will be quicker to strike the back surface of the 'interjected medium' leading to a correspondingly shorter pulse for the aethereal waves.

The implications of this hypothesis were not carried through either in later work or in the manuscript where they appeared, because Newton was influenced by his early reading of his older contemporary, Charleton,[50] into adopting the Aristotelian view that red was the strongest colour because it contained the least 'darkness', meaning that red corpuscles would have the greatest mass and least speed. In the 19th century, Thomas Young drew attention to the fact that Newton's own experiments on thin plates showed that the thickness at which the fringes were produced (and, therefore, the wavelength) decreased in a denser medium, though the corpuscular theory implied that they should increase.[51] This meant that, contrary to Newton's velocity-dispersion model of 1664–1665, light actually moved slower in a denser medium. Remarkably, Young, in his early work, himself retained Newton's Aristotelian views on the strength of red rays, and this informed his belief that the emission theory required a direct proportion between the density of the medium and the thickness of the film.

After 1664–1665, the next time Newton referred to a velocity-dispersion model was when he wrote to Oldenburg in 1676 in reply to a suggestion from Hooke 'that colour may proceed from the different velocities wch aethereall pulses may have as they come immediately from ye Sun.' Here, Newton

---

[49] *Of ye Coloured Circles*, Westfall (1965, 191).
[50] Charleton (1654).
[51] Young (1802b, 393).

described how coloured rays of different velocities could be separated by refraction within his own theory:

> For to suppose different velocities of the rays the principle of colour is only to assign a cause of the different colours which rays are originally disposed to exhibit and do exhibit when separated by different refractions. And though this should be the true essential cause of those different colours yet it hinders not but that the different refrangibility of the rays may be their accidental cause by making a separation of pulses of different swiftness.... so far is this Hypothesis from contradicting me, that if it be supposed it infers all my Theory. For if it be true,...then is there nothing requisite for ye production of colours but a separation of these rays so that ye swiftest may go to one place by themselves & ye slowest to another by themselves, or one sort be stifled & another remain: then must all the phenomena of colours proceed from the separations of these rays of unequal swiftness...Were I to apply this Hypothesis to my notions I would say therefore yt ye slowest pulses being weakest are more easily turned out of ye way by any refracting superficies than ye swiftest, & so *caeteris paribus* are more refracted: and that ye Prism by refracting them more separates them from ye swiftest, and then they being freed from the alloy of one another strike the sense distinctly each with their own motions apart and so beget sensations of colour different from one another and from that which they begat while mixed together; suppose ye swiftest ye strongest colour, red; and ye slowest ye weakest, blew.[52]

He then proceeded to claim (presumably on the basis of his notebook entry) that he himself had 'formerly applied' the idea in 'other Hypotheses', but implied that Hooke's own hypothesis was erroneous because it meant that different colours of light had different vibrations in the aether, while Newton thought it 'much more natural to suppose' that pulses in the aether were 'equally swift and to differ only in bigness, because it is so in ye air, & ye laws of undulation are wthout doubt ye same in aether that they are in air.'[53] This would seem to imply that Newton's 'other Hypotheses' had not assumed that dispersion required different velocities for differently coloured rays 'as they come immediately from ye Sun,' but only when they passed through refracting media.

Newton only ever produced one published derivation of the Cartesian refraction condition for velocity. This was in Section XIV, Book I of the *Principia* (1687), which contains five Propositions 94–98 and is headed: 'Of the motion of very small bodies when agitated by centripetal forces tending

---

[52] Newton to Oldenburg, 15 February 1676, *Corr.* I, 418–419.
[53] *Ibid.*, 419.

to the several parts of any very great body.' Proposition 94 derives Snell's law of refraction, Proposition 95 the Cartesian condition and Proposition 96 explains total internal reflection. There is no mention of light at all until the illustrative Scholium added after Proposition 96 where Newton says: 'These attractions bear a great resemblance to the reflexions and refractions of light made in a given ratio of the secants, as was discovered by *Snellius*; and consequently in a given ratio of the sines, as was exhibited by *Des Cartes*.' At the end of the Scholium, referring to the next two Propositions, he says: 'Therefore because of the analogy there is between the propagation of the rays of light and the motion of bodies, I thought it not amiss to add the following Propositions for optical uses: not at all considering the nature of the rays of light, or inquiring whether they are bodies or not; but only determining the trajectories of bodies which are extremely like the trajectories of the rays.'

Newton is clearly making a point of expressing himself in extremely tentative language, and on this issue he never became more definite. In a manuscript draft of this period, he wrote: 'refraction is already explained similarly . . . whether in fact reflection and refraction are made through attractions he may dispute who wishes.' Bechler, in his analyses of Propositions 94 and 95, has provided the explanation. Descartes' derivation of Snell's law and his refraction condition was based on two assumptions: one was that only the velocity component normal to the surface actually changed; the other was that the ratio v_i/v_r remained a constant for any given material. Assuming that the parallel velocity remains unchanged, however, does not tell us what happens to the vertical component. Newton did not normally make such assumptions without a genuine physical reason and he also wished to explain his own observations of total internal reflection. So, he attempted to explain both the sine law of refraction and the phenomenon of total internal reflection using the same mechanics as he had successfully applied in large-scale systems acting under gravitational forces. This was entirely in accord with the strict 'rule of normality', which, according to Bechler, he applied at this time, with the laws operating in micro-physics acting essentially like small-scale versions of those observed on the large scale.[54]

Making use of Descartes' idea of unchanged horizontal velocity with the vertical force on the particles a function of distance allowed him to develop

---

[54]Bechler (1973, 2–3).

an exact analogy between optical refraction and terrestrial gravity. Total internal reflection, in particular, could be explained by a gradual incurving of the rays near the reflecting surface in exactly the way that a projectile thrown at an oblique angle returned to the Earth's surface under the action of the vertical force of gravity.

According to Bechler's analysis, the argument in the *Principia* implied, without stating, a velocity model of dispersion, but this time the model implied that the refractive index depended on the incident velocity according to the formula

$$\frac{\sin i}{\sin r} = \sqrt{1 + \frac{k}{v_i^2}},$$

where k is a constant of the refracting medium, while the Cartesian condition required that the velocity of a highly refracted ray must be greater than one less refracted, or that blue light would travel faster than red. And, if the refractive index was a function of the incident velocity, the speeds of different coloured light rays would be different in a vacuum as well as in material media.

Newton had a clear opportunity to explain the different refractions of rays of different colour in terms of their different velocities of incidence, but it seems that he elected not to take it, for he was already 'fully assured' that the velocity model, which 'fitted only two of the experimental cases known,' 'was a complete failure' in points of detail. The model in question was certainly *not* a return to the velocity model of the early notebook. The Cartesian condition which Newton derived in Proposition 95 (without using the field of force employed in Proposition 94), predicted that the velocity of a highly refracted ray would be greater than the one less refracted, and that blue light would travel faster than red. In addition, if the refractive index depended on the incident velocity, the speeds of different coloured light rays would be different in a vacuum as well as in material media.

The *Optical Lectures* show that Newton's theory was conceived in terms of momentum, rather than velocity. Though he seems to have always held to the inverse or Cartesian refraction condition for momentum ($n = p_2/p_1$), he came to reject its application to velocity. Newton never committed himself in print to the $n = v_2/v_1$ condition for velocity, implying that it must instead involve a change in *mass*. The only published derivation he ever made was general and not specifically applied to light; and he was careful to say that the application was only a possibility.

4.5. Relativistic Mass and Refraction

That c is implicated in the mass changing on refraction can be seen in Proposition 94 from *Principia*, Book I. Newton's result has been given in analytic form by Shapiro.[55] If we suppose that $f(\rho)$ is the force per unit mass acting on the light corpuscle, with i and r the angles of incidence and refraction, and ρ the distance from the refracting surface, then the refractive index from the denser to the less dense medium becomes:

$$n = \frac{\sin i}{\sin r} = \sqrt{1 - \frac{2\phi}{c^2}},$$

where c is the incident velocity and

$$\phi = \int_0^R f(\rho)d\rho.$$

Now, integrating force over distance (as Newton does in Proposition 41) will give the work done (per unit mass) and hence the kinetic energy gained (per unit mass) as the light corpuscle's velocity component in the direction of the refracting force increases from 0 to, say, v. So, ϕ will become $1/2 v^2$ and 2ϕ (the refracting potential energy per unit mass or 'potential') will become v^2. The refractive index can then be written in the form:

$$n = \frac{\sin i}{\sin r} = \sqrt{1 - \frac{v^2}{c^2}}.$$

For the corpuscular model, $n = u/c$, where u is the (reduced) speed after refraction into the less dense medium. Hence,

$$n = \frac{u}{c} = \sqrt{1 - \frac{v^2}{c^2}}. \tag{1}$$

Another way of writing this is

$$u^2 = c^2 - v^2.$$

This is simply a (Pythagorean) vector addition of velocities u and v to create c, or by implication the vector addition of momenta mu and mv to create total momentum mc. However, (1) also has the structure of the relativistic $u/c = 1/\gamma$, or again, $mu/mc = \gamma$. That is $mu = \gamma mc$, the relativistic mass–momentum decrease of the light corpuscle on refraction into the less dense medium. This may be the first time that a gamma factor appears in physics. The relativistic connection is not so surprising as it first looks, as the

---

[55] Shapiro (2002, 236). See also Chandrasekhar (1995, 325–326).

relativistic equations are really only an application of Pythagorean addition, and 'relativistic mass' is structured to be compatible with classical energy conservation.

In 1850, Foucault completed what was claimed at the time to be a decisive test between the $n = v_1/v_2$ and $n = v_2/v_1$ refraction conditions by using a rapidly rotating mirror to measure the speed of light in both air and water.[56] This was only a year after Fizeau had succeeded for the first time in measuring the speed of light in a terrestrial experiment.[57] Foucault found that the speed was faster in air than in water, so apparently confirming the validity of $n = v_1/v_2$ and refuting the alternative $n = v_2/v_1$. 160 years earlier, however, and before he came to publish his *Opticks*, Newton had actually arranged an experimental test, with very close parallels to that of Foucault; and this led him to reject the $n = v_2/v_1$ condition as a model for dispersion. It forced him to carefully contrive his proof of the sine law of refraction in the *Opticks* in a way which avoided making an explicit statement of $n = v_2/v_1$. His 'oversight' was subsequently corrected by others supplying the missing step.

The test which Newton devised was, in effect, a Foucault experiment using chromatic dispersion. In attempting to provide a physical explanation for the arbitrary assumptions of Descartes' original argument, Newton had derived a formula which suggested that blue rays should travel faster than red over space. Applied to the only system in which it could be tested, the effect would make Jupiter's satellites appear red for a few seconds before they were eclipsed. In a draft for Book II, Part III, Proposition 10 of the *Opticks* Newton writes: 'The most refrangible rays are swiftest. For the light of Jupiters Satellites is red at their immersion,'[58] and in a letter of 10 August 1691 he requests Flamsteed, at Greenwich, to make the observation: 'When you observe eclipses of [Jupiter]'s satellite I should be glad to know if in long telescopes ye satellite immediately before it disappears incline either to red or blew, or become more ruddy or more pale then before.'[59] Significantly, perhaps, Newton apparently also allowed for the non-Cartesian possibility that red light travelled faster than blue.

Flamsteed never found out why Newton had made this request, for Newton had learnt not to reveal his theoretical secrets too readily. However, he made the observation, without finding any indication of the effect,

---

[56] Foucault (1850).
[57] Fizeau (1849).
[58] Draft for Book II, Part III, Proposition 10 of the *Opticks*. Bechler (1973, 22).
[59] *Corr*. III, 164.

which would have lasted for several seconds and so would have been easily noticeable had it existed. The test proved instead to be the strongest possible evidence that the velocity of light was the same for all colours across empty space, an important result in itself. David Gregory noted as late as 4 May 1694 that Newton believed that 'The colour red moves more slowly, blue more quickly: this is inferred from the greater refraction of the red,'[60] a seeming reversal of his earlier position, but Newton had abandoned the velocity model of dispersion by the time he published his *Opticks* in 1704 and an early draft of the suppressed fourth book states that: 'All ye vibrations excited in ye refracting medium by light are equally swift.'[61]

Newton was seemingly obliged to discard the whole theory that he had set out tentatively in the *Principia* but he clearly had no intention of making a public admission of this, especially to Flamsteed, who had little regard for theorists. The test was repeated in the middle of the 18th century without knowledge of Newton's unpublished work. Calculating the speeds associated with the interval of fits rather than the light rays, Courtivron and Melvill both predicted that corpuscles associated with red light would travel faster than those associated with blue. Clairaut developed an exact formula, similar to that defined by Newton in Proposition 94, which predicted a sequence of colours from white through green to violet over a period of half a minute for one of Jupiter's satellites going into eclipse behind the planet, and an opposing sequence from red to white as the satellite emerged from the planet's shadow. The instrument-maker James Short was asked to make the test, and, like Flamsteed, reported negative results.[62] Almost exactly the same kind of evidence — this time involving solar eclipses — helped to turn the young Fresnel towards the wave theory in 1814.[63] As a consequence of the experiment, Newton realised that the relation between velocity and momentum for light was nonlinear. He was unable to explain this but his efforts to do so led to some of the more imaginative speculations of his optical Queries.

The true significance of the Foucault experiment was that it showed that the sine law of refraction was a consequence of the principle of least time, especially when combined with Huygens' application of the wave theory, and not of the principle of least action when applied to ordinary material corpuscles. But we now know that Newton had already shown that

---

[60]'Physical and Mathematical Annotations with Newton', *Corr.* III, 317.

[61]Early draft of suppressed Book IV, Proposition 6 for *Opticks*, 1690s; Bechler (1973, 21).

[62]Courtivron (1752); Melvill (1753); Clairaut (1754); Whittaker (1951–1953, 1: 99); Cantor (1983, 64–66).

[63]Fresnel (1866–1870, 2: 821). The relevant passage is translated and quoted in Buchwald (1989).

the Cartesian condition in terms of velocity did not apply to his kind of light particles, though this did not rule out applying a similar condition to momentum. More than a century and a half before Foucault's test, Newton had made the startling but significant discovery that small-scale forces do not obey the same rules as large-scale forces and that light particles do not behave in the same way as material corpuscles. Ultimately, this came from his 'field' approach to the processes of reflection and refraction. Although the true nature of the 'nonlinearity' of the optic or opto-electric force could not, of course, have been discovered at the time, it is remarkable that Newton should have realised that it was necessary to a successful mathematical derivation of the sine law of refraction from a particle theory.

Whiteside writes of Newton's 'London years, when with increasing age he grew less and less confident that his force-theory of light — with its ineluctable corollary that the speeding light-corpuscle must move faster in a denser medium than a thinner one — accurately models optical reality.'[64] There is no derivation of the Cartesian condition in the *Opticks*, as there surely would have been had Newton still completely believed in it. There is a mention of the condition in Proposition 10 of Book II, Part 3, which was inserted into the printer's copy at the last minute, but this seems to be a way of introducing the sine law of refraction and the explanation which follows is actually given in terms of 'motion', i.e. momentum, where it does apply, and not in terms of velocity.

Newton, however, still required a proof of the sine law of refraction for the *Opticks*, and so he devised one which avoided any mention of the velocity model of dispersion, *or even of the ratio v_r/v_i*; the velocity model would never feature in any of his published works. Though, in view of the earlier difficulties he had faced with his optical theories, Newton would have been very ill-advised to draw attention to this area of apparent failure in his work, he was, as he was writing the *Opticks*, 'all the while filled with the knowledge that the *Principia*'s reduction was wrong in some fundamental aspect'; according to Bechler, 'it was decidedly not in his best interest to remind anybody of it.'[65] He was, of course, under no obligation to 'correct' a model which he had stated in quite explicit terms applied only to forces which may or may not have been like those of light; and he would have been under no illusion that to have done so would have been to give ammunition to those like Leibniz who had attacked the *Principia*'s more certain conclusions in other areas such as gravity.

---

[64]Whiteside (1980, 307).
[65]Bechler (1973, 26); Bechler (1974, 194); *Opticks*, Book II, Part 3, Proposition 8.

Whether or not he drew attention to the failure of his model, it has to be said that this whole investigation was an outstanding example of the scientific method in action. It would have been relatively easy to have come up with some model-dependent hypothesis which covered up the difficulties, or to have simply ignored the unpublished evidence in the published presentation, but Newton knew that science was not well served by this kind of approach. There were several occasions on which his investigations led to an impasse due to apparently irreconcilable elements, but ultimately the recognition of a problem became the key to creating the next stage in the development.

Since the proof in the *Principia*, of course, was only related to light in the most tentative language and in a Scholium separated from the main proposition, and was effectively abandoned with the velocity model, we can no longer maintain that Newton remained a firm believer in the Cartesian condition

$$\frac{\sin i}{\sin r} = \frac{v_r}{v_i},$$

which, it was later claimed, was decisively refuted by Foucault, though he may well have been prepared to maintain the *momentum* ratio p_r/p_i, by speculating that the refractive force was a nonlinear function of mass, an idea associated in his manuscripts with the interconversion of light and matter, and fully compatible with the 'relativistic mass' interpretation of Proposition 94.

4.6. Phase Velocity and the Superluminal

A remarkable idea that developed early on in Newton's optical theory was that the waves associated with light could actually overtake the rays, and cause (in his later terminology) the fits of easy reflection and easy refraction. In effect, these would be *phase waves* travelling faster than the ray itself. Since he uses the analogy of the ripples produced by the waves from a stone dropped in water, and also analysed waves in water in mathematical detail in the *Principia*, it may have been from direct observation of waves produced in this way, where the phase waves would have been clearly visible to such an acute observer.

The idea is fully set out in the *Opticks*:

> But What kind of action or disposition this is; Whether it consist in a circulating or vibrating motion of the Ray, or of the Medium, or something else, I do not here enquire. Those that are averse from assenting to any new Discoveries, but such as they can explain by an Hypothesis, may for

the present suppose, that as Stones falling upon Water put the Water
into an undulating Motion, and all Bodies by percussion excite vibrations
in the Air; so the Rays of Light, by impinging on any refracting or
reflecting surface, excite vibrations in the refracting or reflecting Medium
or Substance, and by exciting them agitate the solid parts of the refracting
or reflecting Body, and by agitating them cause the Body to grow warm or
hot; that the vibrations thus excited are propagated in the refracting or
reflecting Medium or Substance, much after the manner that vibrations
are propagated in the Air for causing Sound, and move faster than the
Rays so as to overtake them; and that when any Ray is in that part of
the vibration which conspires with its Motion, it easily breaks through a
refracting Surface, but when it is in the contrary part of the vibration which
impedes its Motion, it is easily reflected; and, by consequence, that every
Ray is successively disposed to be easily reflected, or easily transmitted, by
every vibration which overtakes it. But whether this hypothesis be true or
false I do not here consider.[66]

Newton is describing a true phenomenon: the 'superluminal' velocity
of the phase wave, the phase velocity of the individual vibrations in a
medium, under certain conditions, being greater than the group velocity
of the modulation produced by the superposition of the individual waves.
Though he describes this in the *Opticks*, where he applies it to the 'fits',
it also occurs much earlier, in the *Hypothesis of Light*, long before the fits
entered into his optical work: 'to explaine the colours, made by *Reflexions* I
must ... suppose, that, though light be unimaginably swift, yet the Æthereall
Vibrations excited by a ray move faster then the ray it self, and so overtake
& outrun it one after another.'[67]

Another version is in the *De vi electrica*, a manuscript of about 1712,
which relates the phase waves (and most other phenomena) to an 'electric
spirit'.

These vibrations are swifter than light itself and successively pursue and
overtake the waves, and in pursuing them they dispose the rays into the
alternate fits of being easily reflected or easily transmitted; and when
excited by the light of the Sun they excite the particles of bodies and by
this excitation heat the bodies; and being reflected at the surfaces of bodies
they turn back within the bodies and persist a very long time.[68]

The superluminal phase waves suggest another reason why he did not
draw the obvious conclusion from his experiments that longer wavelengths

---

[66]Book II, Part 3, Proposition 13.
[67]*Hyp., Corr.* I, 377–378.
[68]*De vi Electrica*, c 1712, *Corr.* V, 365–366.

implied faster rays. Complications in the refraction process were suggesting that momentum could not be simply defined as the product of mass and velocity.

Though it has been noted that the equation

$$\frac{u}{v} = n^2$$

would not have been compatible with Newton's physical description of aether waves overtaking the corpuscles for conditions in which $n > 1$, this process would, of course, occur when $n < 1$.[69] The difficulty arises from the fact that yet another velocity, the signal velocity, is needed to complete the picture. Waves do indeed overtake the light signal (or 'ray' in this sense) in a normally dispersive medium, but then the light signal does not move at the speed of the particles. The confusion arises from the use of 'ray' to mean two quite different things, which are only equivalent under certain particular conditions. Newton knew that particles did not travel at the same speed as aethereal waves in a medium, but, as a result of the nature of refracting and dispersive media, it turns out that the 'speed of light' that we measure is different from either.

In reality, the term 'velocity of light' does not refer to a uniquely defined concept; and the Cartesian condition, the Hobbesian condition and Foucault's measurements are concerned with three entirely different physical quantities. The Cartesian condition is, as Newton thought, applicable to the photon momentum. If we divide it by the 'relativistic' mass, we can, in principle, define a 'particle velocity' (p/m). This would only be meaningful in a medium or under conditions where the photon was effectively subjected to a force and the γ factor invoked. The Hobbesian condition applies to the phase velocity, the velocity of individual waves ($v = \nu\lambda$, using Newton's formula), which in a dispersive medium will vary significantly over a wide range, so producing the chromatic dispersion. It can be either subluminal or superluminal, depending on whether the dispersion is 'normal' or 'anomalous'. If ω is the angular frequency ($2\pi\nu$) and k the wave number, then the phase velocity is given by $d\omega/dk$. However, what is normally measured in an experiment like Foucault's is the group velocity (v_g) or the velocity of a group of waves. The group velocity, which is defined by ω/k,

---

[69]Stuewer (1969, 393–394); $n < 1$ would be a case of the so-called 'anomalous dispersion' in which the spectral colours are reversed. It can be seen, in a relative sense, in a hollow prism immersed in a water medium. Though such an arrangement would have been available to Newton (Aboites, 2003), there is no evidence that he exploited it to make this observation.

can be close, in certain circumstances, to the phase velocity, but in others it approximates even more closely to the 'particle velocity', and in many cases it is quite different from either. It can even be in the opposite direction, leading to a 'negative' refractive index. In Newtonian theory, this can perhaps be equated to the velocity of the light rays, seen as modulations. The whole idea of a clash of opposing 17th century theories being resolved by the superior experimental techniques of the 19th century is, in fact, largely an artefact of the latter century.

All three quantities referred to by the label 'velocity of light' are the same for light in a vacuum (c) but are very different for light in a medium. Both the Cartesian and Hobbesian conditions make entirely correct predictions for the behaviour of their quantities and their results are reconciled by de Broglie's relation which also reconciles the principles of least time and least action. Thus,

$$p = \frac{h}{\lambda}$$

becomes

$$u = \frac{c^2}{v},$$

where c is an absolute constant. Wave and particles velocities *are* inverses of each other. The first is increased in passing into an optical denser medium (or one of higher refractive index), while the latter is decreased. But *neither* set of predictions can be tested by the experiment devised by Foucault, which measures something completely different, except in extremely favourable circumstances! The group velocity, which is the usual quantity measured by an experiment like Foucault's, is generally decreased in a physically denser medium though this may not necessarily be optically denser. The differential equation which determines the relation between the wave and group velocities indicates that a given group velocity does not uniquely determine the component phase or wave velocities.

According to measurements of radiation pressure, photons do increase in momentum at the boundary of a denser medium exactly as predicted by the Cartesian condition. As Jones and Leslie, were able to demonstrate in 1978: 'the momentum associated with electromagnetic radiation increases directly with the refractive index of the medium into which it passes, discriminating substantially in favour of the phase velocity ratio and against the group velocity ratio.' According to Jones: 'This simple result had been expected (Jones 1951) on the naive arguments of Newtonian corpuscles . . . and of the relation $p = h/\lambda$ for a photon.'[70] It is of additional interest in this story that, in relativistic quantum mechanics, nothing travels at the speed of light because the correlations between different quantum systems are instantaneous (via vacuum and perhaps gravity); and that, in another sense, everything travels at c, because all material particles travel at exactly this speed, though an oscillating motion or *zitterbewegung* between ordinary space and vacuum, which occurs at a frequency determined by the mass, effectively makes the motion appear to be subluminal.

Several other results of the 20th century seem to point in the same direction as those of Jones and Leslie. Poynting and Barlow showed a classical momentum increase in 1904, the theory of their experiment being based on Larmor's general application of the concept of momentum to a wave train 'without any further appeal to the theory of wave motion,'[71] in effect, a recognition that specifically wave aspects of the theory need not be invoked. Thomson had, a little earlier, used the wave theory of light to show that the momentum increase produced by a change of medium was the same as had been predicted by the corpuscular theory.[72] A change of photon mass on refraction (from $h\nu/c^2$ to $n^2h\nu/c^2$), which had been hinted at by Thomas Preston in his textbook on *The Theory of Light* in 1895, was established on a mathematical basis by Michels and Patterson in 1941.[73]

However, despite these results, complete resolution of the problem was only achieved in 2010 when it was established by Barnett and Loudon[74] that the momentum which increases is the *canonical* momentum ($nh\nu/c$), the momentum transferred to a body in the refracting medium, as opposed to the *kinetic* momentum ($h\nu/nc$), the momentum not transferred, which decreases, and which follows the phase velocity ratio, as other (though perhaps less decisive) experiments had seemed to indicate.[75] In the early years of the quantum theory, the two conditions were championed, respectively, by Minkowski and Abraham, and led to a century-long controversy, which turned out to be another manifestation of wave-particle duality, with an identical

---

[70] Jones and Leslie (1978). See also Jones and Richards (1954).

[71] Poynting (1905b).

[72] Thomson (1903).

[73] Preston (1895, 8, footnote 1 (not in first edition of 1890)). Tangherlini (1961, 1975) first drew attention to Preston's result. Michels and Patterson (1941).

[74] Barnett (2010), Barnett and Loudon (2010). Appropriately, the Barnett and Loudon paper was published in the *Philosophical Transactions*, 338 years after Newton's first optical paper had appeared in the same journal.

[75] Walker *et al.* (1975).

resolution.[76] The distinction between canonical and kinetic momentum only has a meaning for a force which is relativistic, like electromagnetism, where, for a charged particle, the vector potential **A** adds an additional $q\mathbf{A}$ to the kinetic definition of momentum as $m\mathbf{v}$.

Essentially, the canonical momentum is really the one demanded by quantum phenomena and by wave effects such as diffraction. It is the momentum in the de Broglie relation and in Schrödinger's quantum prescription $\mathbf{p} = -i\hbar\nabla$, and its refractive index is determined by dividing c by the phase velocity. In a refracting medium, the effect is produced by the entire field within the medium, as Newton, in a sense, realised, and not by the kinetics of a single particle. In fact, in such a medium, the field quanta are not pure photons, but polaritons, produced by coupling the photons with excitation states within the material.[77] The kinetic momentum, however, emerges in Abraham's theory from a semi-classical treatment of the kinetics of an isolated photon, and, in a dispersive medium, it requires the refractive index to be determined by the group velocity, not the phase velocity. It is most effective in media where these are closely related. It is remarkable that Newton's work in optics required a concept of momentum for light in a medium which went beyond the simple kinetic definition as 'mass $\times$ velocity' because light particles required a new understanding of the concept of 'mass' and its relation to force.

The complicated story of the velocity of light, and, in particular, Newton's contribution suggests that over-reliance on the 'hypothetico-deductive' model of scientific reasoning causes problems because it assumes that there is such a thing as an isolated hypothesis which can be tested on the grounds of its own predictions. However, scientific predictions cannot be derived solely from single hypotheses but are constructs derived from a combination of hypotheses and always at several stages removed from any single one of them. It is nearly always possible to make the same predictions from opposing hypotheses if the right subsidiary hypotheses are selected. Thus, the alternative refraction conditions of Descartes and Hobbes could each have been derived from either particle or wave hypotheses — and were. Foucault's classic experiment, even if it had been really decisive, would not have established whether light was particle- or wavelike, but whether the theory of light as ordinary material corpuscles subject to the principle of least action was a more plausible description than the theory of light as a wave motion subject to the principle of least time. The element in the

---

[76]Minkowski (1908); Abraham (1909, 1910).
[77]Barnett (2010); Barnett and Loudon (2010); citing Kittel (1987).

particle theory which was finally rejected was neither the corpuscular nature nor the principle of least action but the idea that light could be considered an ordinary material body, an idea which Newton himself had already been forced to question. Testing and prediction, of course, are very important processes in the development of scientific theories, but they should never be given primacy over the more fundamental principle of simplicity, especially when the predictions are based on extensive mathematical analyses.

Chapter 5

Mass–Energy

5.1. $E = mc^2$ and Classical Potential Energy

The famous equation $E = mc^2$ did not arise spontaneously from Einstein's special theory of relativity as popular accounts often imply. In fact, it had a long line of antecedents and, like the theory itself, represented a culmination of a tradition rather than a revolutionary new beginning. Certainly, it represents the equivalence of mass and energy, perhaps for the first time, in a clear-cut and unequivocal manner, but something like this equivalence had long been suspected by many previous writers, especially the late-19th century physicists who were working out the detailed consequences of Maxwell's electromagnetic equations. What is particularly remarkable about this equation is how little it actually has to do with the theory of relativity.

In fact, the relationship $E = mc^2$ cannot be derived from special relativity by any deductive process, as Einstein found out after much effort over many years. The standard derivation begins with the Lorentz-covariant equation between the rate of increase in kinetic energy dT/dv and the rate of doing work, $\mathbf{F} \cdot \mathbf{v}$.

$$\frac{dT}{dt} = \mathbf{F} \cdot \mathbf{v}.$$

This is then integrated to find the total kinetic energy of the system

$$T = \frac{mc^2}{\sqrt{1 - v^2/c^2}},$$

which expands to give

$$T = mc^2 + \frac{mv^2}{2} + \cdots .$$

133

It is this equation which defines the 'rest energy' of the system (or its total energy when in a state of rest) to be mc^2, which is then added to the classical $mv^2/2$, and any higher order terms, to give the total relativistic energy.

However, this is only possible if we assume 0 as a constant of integration. As is well known, constants of integration are arbitrary unless specific boundary conditions are known, and in this case they are not. So, we have no justification from within the system for taking mc^2 as the rest energy. If we took the constant of integration to be $- mc^2$, for example, then the rest energy would be 0, and the total energy would become $mv^2/2$, for $v \ll c$, exactly as classical conditions require. Of course, we can always define *changes* of energy (ΔE) as equivalent to *changes* in mass (Δmc^2), which ensures that $\Delta E = \Delta mc^2$ is a totally valid deduction from the special relativistic postulates, but this doesn't fix the starting-point at $E = mc^2$. In all versions of the theory, this is an assumption made only for 'convenience'.

Writers who have closely analysed Einstein's own arguments find that, though the expression $\Delta E = \Delta mc^2$ follows without difficulty from his basic assumptions, any extension to defining the rest energy as $E = mc^2$ will necessarily be arbitrary and non-deductive. For Stachel and Toretti, for example: 'The final conclusion that the entire mass of a body is in effect a measure of its energy, is of course entirely unwarranted by Einstein's premises.'[1] But Einstein makes the same point himself: 'A mass μ is equivalent — insofar as its inertia is concerned — to an energy content of magnitude μc^2. Since we can arbitrarily fix the zero (the rest energy), we are not even able to distinguish, without arbitrariness, between a "true" and an "apparent" mass of the system. It appears much more natural to regard all material mass as a store of energy.'[2]

The question then would become: why is it convenient? And the answer would have to be that $E = mc^2$ is a fundamental truth which is not part of special relativity itself, but one which that theory has somehow uncovered. Such things are not unusual in physics, and especially in connection with relativity. Other examples include the Schwarzschild radius for black holes, the equations for the expanding universe, and the gravitational redshift, but a particularly striking one is the derivation in quantum mechanics of the spin $1/2$ state of the electron from the relativistic Dirac equation. The spin $1/2$ condition has nothing to do with the fact that the equation is relativistic or 4-vector in space and time or energy and momentum. It comes entirely from the vector properties of the momentum and space

---

[1] Stachel and Toretti (1982, 761).
[2] Einstein (1907, 442); quoted Stachel and Toretti (1982, 761).

terms. But, to make the equation fully relativistic, it becomes necessary to be more specific about the exact algebra connected with the vector terms, and the extra information required turns out to be the source of the spin $1/2$ condition. Once this is known, it is then possible to derive the same condition by applying the same extra information to the vector terms in the entirely non-relativistic Schrödinger equation.[3] The extra information was, of course, always there even before the physics became relativistic, but making it relativistic required changes which forced it out into the open.

The extra information which relativity forces onto us in the case of the relativistic mass–energy derivation comes from the assumption that it is convenient to preserve the classical laws of conservation of mass, energy and momentum even under relativistic conditions, and this tells us something that we previously didn't know about classical energy conservation. $E = mc^2$ is, in fact, a *classical* energy condition which has to be incorporated into relativity to make the classical-relativistic connection work. It describes a classical potential energy intrinsic to the concept of inertia, but not normally detectable because under normal classical conditions it is unchanged. This is exactly the way we would expect potential energy to work, and the expression mc^2 has the precise mathematical form we would expect such a potential energy to have, without the factor $1/2$ that would distinguish it as kinetic.

An intrinsically arbitrary nature of the numerical value is typical of potential energy terms. Of course, though it may be arbitrary in special relativity, the value need not be arbitrary in other contexts. An idea which is 'physically reasonable' or 'natural' should be explicable in terms of some definite physical principles; and if mass is to be considered as a 'store' of energy, then this principle must be related to the idea of mass as a specifically *potential* form of energy. As a result of the fundamental condition of gauge invariance, we can only detect *differences* in potential energies, not their absolute values, if these have any meaning. If no differences occur (for example, when the material nature of a body determining its rest mass is preserved in a context where $v \ll c$), then we will not detect anything. But, looking from a different perspective, as in relativity, and trying to reconcile it with the original one, we may well expose the hidden assumptions that were made in the latter case.

The correctness of the identification of mc^2 as an expression for *potential* energy is shown by the classical derivation, by Boltzmann in 1884, of Stefan's fourth-power radiation law from a 'radiation gas', entirely analogous to the

---

[3]This development is based on work initiated by Hestenes (1966). A version of the calculation for the Schrödinger equation can be found in Gough (1990).

gas of material particles assumed in classical kinetic theory.[4] While the potential energy equivalent for a material gas is derived from the product of pressure and volume, and is twice the total kinetic energy of the molecules moving in any given direction and is $mc^2/3$ for each degree of freedom, the energy of the radiation gas is the product of the radiation pressure and volume and is also $mc^2/3$ for each degree of freedom. The expression mc^2 in this context clearly does not depend in any way on the 4-vector or 'relativistic' properties of light, for the analogous system of gaseous molecules is not 4-vector or relativistic.

In fact, it is now possible to see that a redefinition of mc^2 as a classical potential term solves an anomaly in relativistic physics, for the transition from special relativistic to classical mechanics at low speeds does not follow unless mc^2 *also has a classical meaning*; and it additionally explains why the mass–energy equation does not arise as a deductive consequence of the special theory of relativity, but must be derived *inductively* from the special relativistic equation $\Delta E = \Delta mc^2$. In these terms, $E = mc^2$ is a classical equation and its justification in relativistic physics is that it represents the classical limit to which relativistic mechanics converges under appropriate conditions. Relativistic results which coincide with classical ones must necessarily be classical in origin, even if it takes relativistic reasoning to uncover them. The relationship between the conservation laws of mass, momentum and energy, in fact, is nothing to do with relativity as such, but is rather a hidden assumption of Newtonian physics. And relativity introduces the term mc^2 to make its energy equations compatible with the Newtonian ones concerned with energy.

5.2. Newtonian Insights

It could, of course, be argued that the late-19th century authors who came close to defining E as mc^2 on the basis of Maxwell's equations (for example, J. J. Thomson, Oliver Heaviside and Henri Poincaré) were using a version of 'relativity before relativity,' and so their derivations don't take us much further in explaining the classical meaning of the term. For this, we need to go back to a much earlier period before electromagnetic theory was developed, and, as in so many other cases, Newton provides us with the key insight.

Query 30 of the *Opticks* poses a remarkable question, which has drawn a great deal of subsequent comment: 'Are not gross Bodies and Light convertible into one another, and may not Bodies receive much of their

---

[4]Boltzmann (1884).

Activity from the Particles of Light which enter their Composition? For all fix'd Bodies being heated emit Light so long as they continue sufficiently hot, and Light mutually stops in Bodies as often as its Rays strike upon their Parts...' The same Query also asserts that: 'The changing of Bodies into Light, and Light into Bodies, is very conformable to the Course of Nature, which seems to be delighted with Transmutations.' Statements like this in Newton usually have an ultimately alchemical origin, which seems to be confirmed by further discussion in the Query of such transformations as that of water into earth, which he had taken from Boyle and ultimately van Helmont. However, by 1706, the 'rationalising' process ascribed to Newton's alchemical ideas had long been established, bringing much of the work into line with material derived directly from physical observation and reasoning.

If we look only at the published text, we might think that Query 30 is just a brilliant speculation with a mostly fortuitous connection with 20th century ideas on the interconversion of energy and matter, but the manuscript material from which the Query was composed tells us a very different story — one that involves the development of a line of mathematical reasoning with a clear connection to the modern derivation of $E = mc^2$. This development also gives us an idea of how we can solve the anomaly in the modern theory, which makes relativity unable to derive the relationship between mass and energy, but only the relationship between their changes.

Query 30 of the English edition of the *Opticks*, published in 1717, was originally Query 22 of the Latin edition of 1706. The drafts for this Query introduce the idea that the refraction of light at a plane boundary is due to a centripetal force mc^2/r acting over distance r, in line with Newton's view that all forces are centripetal forces, and, under steady-state conditions, a force of this form, is exactly equivalent to a potential energy of the form mc^2. It is important to note that there is no physical model involved in this argument. It is purely an abstract relation between fundamental physical quantities, essentially a form of dimensional analysis, which would be true irrespective of the physical mechanism involved in the process.

According to this reasoning, the expression mc^2 for the electromagnetic energy of photons or for the intrinsic energy of Newton's light corpuscles comes from the fact that it is a potential energy term, which, by application of the virial theorem, takes a value equivalent to twice the kinetic energy in any steady-state system under the action of either constant or inverse square law forces, although for a constant force it is positive and for an inverse square force negative. Exactly such a relation between kinetic and potential energies (though sometimes with a different numerical factor) emerges in all steady-state systems regardless of the physical nature of the force or type of potential which is assumed to operate, and, with potential energy replaced by the work done, it applies (in a more general form) to practically any system not subject to dissipative forces.

The difference in numerical value between the potential energy or total work done in such a system and the normal expression for the kinetic energy induced when work is done in changing the speed of a body is due solely to the steady-state conditions which apply in the former instance, and it is notable that the *half*-value $^1/_2\, mc^2$ is used as the energy term in cases where systems are described in specifically *kinetic* terms. An example occurs in the derivation of the Schwarzschild radius, where c is taken as the maximum speed of a particle escaping from a gravitating body. The potential nature of mass–energy should be perhaps obvious from its incorporation of the idea of 'rest' energy; energy due to matter at rest is not, in the normal sense, 'kinetic'. Even photon energy can be described as 'potential' in this sense, for the photon only has existence as a 'particle' at the point of interaction.

The idea of mc^2 as a potential energy for particles of light is also implicit in Newtonian corpuscular theory, for example, in the calculations of atmospheric refraction and refraction at a plane boundary which lay behind the apparent speculations in Newton's Query 30. It is also apparent in the 18th century calculations of gravitational light bending, and again, as we have seen, in Boltzmann's 19th century gas-analogy for radiation pressure. The term mc^2, or its verbal equivalent, actually occurs in Newton's early work in optics. The essay *Of Colours*, which is thought to date from about 1666, contains a calculation of the enlargement to be expected in the area of the interference rings produced in a thin film when viewed through an oblique angle.[5] Here, the 'factus of ye motion of ye incident ray into its velocity,' which is the measure of the strength of the ray, is the product of the momentum ($p = mc$) and the perpendicular component of the velocity ($c \cos i$), and, when the ray is perpendicular, is precisely mc^2. The areas of the rings, in Newton's formula, were inversely proportional to the energies due to the components of their motions perpendicular to the film.

Newton's use of the term mc^2/r as the force exerted on light particles during refraction is entirely consistent with the use of mc^2 as the potential energy of the system, in the same way as mv^2 is the potential energy for bodies kept in gravitational orbit by the centripetal force mv^2/r. The constancy of the velocity of light ensures that the optical system is 'steady-state' and that its potential energy term is numerically equivalent to that

---

[5] *Of Colours*, 1666, Chemistry Notebook, McGuire and Tamny (1983, 466–489), quoted Westfall (1965, 188).

in the inverse-square-law gravitational system. The gravitational case is a particular example of the virial theorem with which Newton was certainly familiar. Another is his formula for the velocity of waves in a medium, in terms of elasticity or pressure (E) and density (ρ), which he applied to both light and sound, Essentially, Newton's formula

$$c = (E/\rho)^{1/2}$$

is an expression of the fact that the potential energy of the system of photons or gas molecules (mc^2) is equal to the work done at constant pressure as product of pressure and volume.

The formula as applied to light occurs in Query 21 when Newton calculates the ratio of the elasticity of a newly proposed electro-optic 'aether' to that of atmospheric air; surviving manuscripts show that this calculation was linked with that of the force of refraction. The elasticity of the aether in Newton's calculation is what we would call energy density of radiation (ρc^2), and this is related by a formula derived by Maxwell in 1873 to the radiation pressure. So, the ratio Newton calculates can be described as the ratio of the energy per unit mass of a particle of light to the energy per unit mass of an air molecule, as manifested in the transmission of sound. Now, the molecular potential energy per unit mass for air may be calculated from the kinetic theory of gases or the equation of state for an ideal gas ($PV = RT$) at about 1.8×10^5 J kg^{-1} when $T = 300$ K. Since the energy per unit mass for a light photon is 9×10^{16} J kg^{-1}, the ratio is 5×10^{11}, which is comparable with Newton's 4.9×10^{11} minimum in Query 21. (Though the calculation is only approximate in principle, there is a relatively close numerical agreement because the error in making the velocity squared ratio of light to sound too low by a factor of 1.5 is compensated for by the factor of 1.4 which is introduced by taking the velocity of sound as the molecular velocity.) The ratio has a special significance in being the potential energy per unit mass ratio characteristic of the two highest levels of organisation in matter, the interatomic and the intermolecular, and it relates to Newton's observations elsewhere that a higher level of force or energy is required to expel light particles in chemical reactions than to expel molecules from liquids as gases and vapours.

The Newtonian formula for calculating the velocities of waves in a medium is a perfect illustration of an application of the virial theorem, which is why it works in the case of light waves travelling in a vacuum, where the model of interaction with matter no longer applies. The centripetal force calculation used in Newton's studies of refraction is yet another such illustration, and will give the correct numerical energy relation whether or not the description of the force is correct. The constraints which need to be applied to find the true nature of the vector force are not required to find the numerical value of the scalar energy.

Newton clearly knew exactly what he was doing in these calculations. There are several theorems in the *Principia* showing how the same energy equations can be applied to many different types of force condition. Light, of course, will give the 'correct' results when travelling through a vacuum, because in these circumstances there is no source of dissipation, and the virial relation takes on its ideal form. The elasticity of light then becomes precisely the same thing as its energy density. Newton was aware that sound in air did not give such perfect results. This, as we now know — though he did not — is because the relatively slow compression and rarefaction of the gas prevents the system from maintaining thermal equilibrium. Laplace, of course, was later to establish agreement between theory and experiment, even in this case, by taking the elasticity at the adiabatic value rather than the isothermal.

In fact, the expression $E = mc^2$ and the ideal form of the virial theorem are truths which are more fundamental than dynamics itself, on the same level as the conservation of mass and charge, or the relationship between such conservation laws and fundamental symmetries. The validity of such archetypal truths is independent of the context in which they were first discovered, and they are no more restricted to the terms of some particular physical theory than the principle we call the 'second law of thermodynamics' is restricted to the terms of the 19th-century thermodynamics of heat engines. We may, therefore, expect to find them in 'classical', as well as in relativistic, contexts. Special relativity certainly points to the universality of the mass–energy relation and extends it from the particular case of the light quantum to all other types of matter, but it is not the ultimate source of the relation and cannot be used to derive it.

The sheer generalising power of the virial theorem is shown by the way it linked several of Newton's separate investigations (although probably not through any conscious realisation on his part) and allowed him to develop unexpected mathematical analogies between apparently unrelated phenomena. In Newtonian physics the centripetal force law mv^2/r is closely linked with systems acting under the force of gravity, and so he developed his calculations on the force of refraction by comparison with such systems. Significantly, Newton's earliest derivation of the centripetal force law involved essentially the same collision process (involving aether particles) as is now used in the kinetic theory of gases. The force was calculated as the product of the change in momentum in a particle due to impact and the rate at which collisions took place, the collision rate being found by dividing the particle velocity by the distance travelled between collisions. Newton subsequently derived it, without the impact mechanism, from the geometry of the circle alone, and he used similarly abstract arguments when he showed that an inverse square law of attraction was the only physically probable law which would explain the observed orbits of the planets.

5.3. The Gravitational Deflection of Light

As a result of his own experimental work in optics, Newton was led to regard the processes of reflection and refraction by mechanisms which in modern terms would be described as *field*-induced effects: 'the Reflexion of a Ray', he wrote, 'is effected, not by a single point of the reflecting Body, but by some power of the Body which is evenly diffused all over its Surface, and by which it acts upon the Ray without immediate Contact.'[6] The same must also apply to refraction, for 'Bodies reflect and refract light by one and the same power, variously exercised in various Circumstances.' The application of a field approach in Newton's refraction calculations made this work directly comparable with classical energy calculations based on the summation of scalar potentials, and, together with the property of periodicity, it led eventually to a recognition of the nonlinearities in the optical force which Newton observed but was unable fully to explain.

Though the reflecting and refracting power was originally conceived in terms of a Cartesian mechanical aether, the aether concept was abandoned during the composition of the *Principia* (1684–1687), in favour of forces acting at a distance. So, in reinterpreting his optical work of the 1666s and 1670s in manuscripts of the 1690s and 1700s, Newton worked by analogy with the gravitational force at the Earth's surface, which, as the collective effect of the Earth's individual particles, he was already conceiving in terms of a field principle.

As a result of his belief in the analogy of nature, Newton pursued an optical/gravity analogy in the process of refraction, in which bodies act on light at a distance to bend the rays, in the same way as bodies act gravitationally on other material bodies to deflect them from a straight-line path. In his very early philosophical notebook, he had speculated on gravitational 'radiation' having characteristics like those of light rays, asking 'whither the rays of gravity may bee stopped by refecting or refracting

---

[6] *Opticks*, 1706/17, addendum to Book II, Part 3, Proposition 8.

them.'[7] Later calculations introduced a direct analogy between gravitational and optical action, and, although Newton tended towards a particle theory of light, others later considered the possibility of gravitational waves on the basis of optical or electromagnetic analogies.[8] Query 1 of the *Opticks* was explicit in connecting gravitational action with light: 'Do not Bodies act upon Light at a distance, and by their action bend its Rays; and is not this action (*cæteris paribus*) strongest at the least distance?' While this certainly referred to the short-range force involved in optical refraction, it was also meant to include the longer-range action of gravitation, especially as Newton regarded light as a material body.

Now, while the comparison between the calculations for pressure and velocity in material and radiation gases gives the clearest indication that the term mc^2 in quantum theory is exactly equivalent to the potential energy term mv^2 in classical physics, there is also direct evidence for the analogy in Newton's own preferred instance of centripetal forces acting on light rays. This time, the calculation, though definitely Newtonian in principle, actually postdates him. However, the independence of $E = mc^2$ from the more general assumptions of relativity theory is demonstrated in a very striking way by these classical or 'Newtonian' derivations of the gravitational bending of a light ray.

Two late-18th century calculations predicted a gravitational deflection for a light ray from a distant object passing a large astronomical body such as the Sun. These calculations effectively assumed that the exceptional speed of the light ray required a hyperbolic orbit of eccentricity e, and distance of closest approach r to a body of mass M, while making an angle ϕ with the horizontal axis. In an unpublished note of about 1784, the reclusive Henry Cavendish, inspired by his friend John Michell's prediction of astronomical bodies with escape velocities greater than that of light, similar to what we now describe as black holes, wrote: '*To find the bending of a ray of light which passes near the surface of any body by the attraction of that body...*'

'Let s be the center of a body and a a point of surface. Let the velocity of body revolving in a circle at a distance as from the body be to the velocity of light as $1:u$, then will the sine of half bending of the ray be equal to $1/(1+u^2)$.' Cavendish's calculation (in which u^2 is presumed to equal rc^2/GM) assumed a velocity of a light particle that was u at infinity, with $\cos\phi = -1/e$. It was

---

[7]QQP, *Of Attomes*, 121^{v}, NP.

[8]One of the most significant of these was Heaviside's pre-relativistic derivation of gravitational waves from gravitomagnetic Maxwell-type equations (1893a, 1893b, 1894) considerably before either form of relativity.

based on a typically Newtonian-type calculation for a hyperbolic orbit of eccentricity e in which

$$c^2 = GM(e-1)/r.$$

A slightly different calculation by the astronomer Johann von Soldner, which was published in an astronomy year book for 1801,[9] was similarly inspired by a 'black hole' prediction, this time by Laplace. Soldner assumed that the light particle velocity was c at the distance of closest approach, with

$$c^2 = GM(e+1)/r.$$

With $e \gg 1$, both calculations predict that the light ray will experience an angular deflection of $2\delta \approx 1/e \approx 2GM/rc^2$. For a light particle of mass m, this is precisely the same result as would be obtained from a potential energy equal to mc^2.

As is well known, this is only half the total deflection, and it has been attributed in relativity specifically to the effect of time dilation or gravitational redshift; Einstein derived it in this way in 1911 using special relativity and the principle of equivalence, and elsewhere he used the same assumptions in calculating the redshift.[10] But there is no need, in fact, to use either the principle of equivalence or special relativity to derive the gravitational redshift. We need only use the quantum theory and $E = mc^2$, and to first order we need only assume that a quantum of energy $h\nu$ and mass $h\nu/c^2$ is subjected to a gravitational potential $-GM/r$ to produce the redshifted $h\nu \, (1- GM/c^2r)$. The quantum nature of this energy ensures that this becomes equivalent to a time dilation which, to first order, is the same as the relativistic $\gamma = (1- GM/c^2r)^{-1/2}$ and this, in general relativity, is the term responsible for the 'classical' half of the light bending.

From various attempts at deriving the full bending of light and the related effect of planetary perihelion precession, it has become clear that the principle of equivalence, is a phenomenon which only has a local application, and only a heuristic value for incorporating relativity or 4-vector theory into physics. If we compare classical and relativistic results we find that the former give exactly what we would expect by assuming $E = mc^2$ without incorporating the specific 4-vector mechanism or relativistic space–time connection. Einstein's 1911 calculation turns out to be effectively non-relativistic.

---

[9] Cavendish (c 1784/1921); von Soldner (1801/1978); Will (1988).
[10] Einstein (1911).

The classical calculation of light bending shows that mc^2 is a potential energy term in classical physics which has the same effect as the equation $E = mc^2$ in relativistic physics, and that all the effects which depend only on $E = mc^2$ and not specifically on the 4-vector combination of space and time can be derived by classical approaches entirely independent of any concept of relativity. The all-embracing nature of the general relativistic formalism has tended to obscure the true nature of the contributions made by different causes to the three relativistic predictions of redshift, light bending and perihelion precession, for none of these effects requires the full field equations of general relativity, as is sometimes assumed.[11] A comparison with classical predictions demonstrates fairly easily that redshift and the time dilation components of light bending and perihelion precession depend only on the relation $E = mc^2$ and not on the 4-vector combination of space and time. The spatial components of the light bending and perihelion precession then follow automatically from the application of 4-vector space–time without any need to apply the equivalence principle, any time dilation necessarily requiring an equivalent length contraction.

There is, however, yet another twist in the story. Though we don't know what formula Cavendish used, we do know that he was inspired by a calculation based on escape velocity, and that Soldner based his argument directly on a similar calculation. In other words, he was effectively using the kinetic energy rather than potential energy equation, Newton's Proposition 39 from *Principia*, Book I, rather than Proposition 41. This is exactly what you would expect for a light ray *forced into* an orbit, as here, rather than already in orbit, and it would be equivalent to

$$\tfrac{1}{2}c^2 = GM(e - 1)/r.$$

This formula gives the complete relativistic light-bending, and would presumably have given Soldner the correct result if he had been able to find the complete integral for the double angle from the hyperbolic function. For light travelling in a straight-line path, the energy is entirely the potential value mc^2, but when it is deviated from its path by a refracting force, then kinetic energy becomes meaningful. Contrary to many statements, a

---

[11]A perennial theme in the literature is that only general relativistic calculations of these effects are 'really' valid, and that the resemblance of the equations produced to similar ones derived using special relativity or classical physics is fortuitous or coincidental. Such coincidences do not occur in physics. If a theory like general relativity actually *incorporates* ones like special relativity and Newtonian gravity as essential components, and needs them to work out its physical meaning, then it should be *expected* to produce results that also stem directly from these theories.

calculation that we can recognise as purely Newtonian gives the complete general relativistic deflection of light.[12]

If there is no known calculation of the gravitational bending of light by Newton, there is a calculation of atmospheric refraction which is similar in principle.[13] Here, Newton investigates the effect of a constant refracting field f at a height h above the Earth's surface, entirely analogous to the gravitational field g $(= GM/r^2)$. He then uses conservation of energy, in the form of Proposition 41, to calculate the resulting deflection into hyperbolic orbits of light rays entering the Earth's atmosphere. The assumption of hyperbolic orbits requires mc^2 to be equated to the potential energy term mfr $(1 + \cos \phi)$, equivalent to the gravitational GMm $(1 + \cos \phi)/r$, while the use of Proposition 41 is equivalent to a modification of c^2 by the factor $(1 - 2fh/c^2)$ in the same way as the principle of equivalence is used to modify c^2 by $(1 - 2gr/c^2)$ or γ^{-2} in gravitational bending. In relativity this modification makes the term $c^2(dt)^2$ into $\gamma^{-2}c^2$ $(dt)^2$ which, with c assumed constant, is interpreted as causing a dilation in measured values of time. It is significant also that, if we apply Newton's atmospheric refraction calculation for a constant refracting field to gravity in a constant field, like that at the Earth's surface, we get a momentum ratio, which is the same equation as that for gravitational redshift. In addition, another of the tests of general relativity, the time delay of an electromagnetic signal, say from a star, in a gravitational field like that of the Sun, is calculable entirely from the refractive index formula:

$$c_{\text{star}}/c_{\text{Sun}} = (1 - 2GM/Rc_{\text{star}}^2)^{-1/2}.$$

Atmospheric refraction was one of the first successes of the corpuscular theory. It could also be said to be one of the longest lasting, as it is still used *in its original form* for practical calculations in spherical astronomy. The classical calculation, initiated by Newton, works in practice because atmospheric refraction effectively assumes a constant speed for the light particles. Sensitive to the dangers of exposing calculations that did not

---

[12]We shouldn't regard the fact that a classical equation gives the correct answer as a 'coincidence'. We should rather ask why it is *this particular* equation. This is where special relativity comes in. In the case of light, c is not a dynamic velocity until the light photons come under a field of force and are deflected. This is why the classical kinetic equation is valid in this instance and not the potential. The classical equation is true, but the full explanation also requires special relativity.

[13]Newton to Flamsteed, 17 November 1694, *Corr.* IV, 46–49; 4 December, *ibid.*, 52–53; 20 December, *ibid.*, 61–62; 15 January, *ibid.*, 67; 26 January 1695, *ibid.*, 73–75, 16 February, *ibid.*, 86–88; 15 March, *ibid.*, 93–97; 23 April, *ibid.*, 105–109.

produce tidy solutions, as this did not when all the contributing factors were taken into account. Newton shrouded his work in this area in secrecy, but he allowed his calculations to stand in the tables published by Halley in 1721, without supplying the theoretical justification.[14] Biot brilliantly recovered Newton's argument in the 19th century,[15] but D. T. Whiteside and particularly Waldemar H. Lehn have greatly expanded our understanding in later years.

Newton, as we have seen, began his calculations in the autumn of 1694 by assuming a constant refracting field within the Earth's atmosphere, similar to the constant gravitational field acting on falling bodies; this was equivalent to a constant density gradient, or variation of atmospheric density from the Earth's centre. Already in a manuscript of the mid-1680s he had written: 'The attraction of ye Ray is proportional to ye variation of density.'[16] He now applied the 'conservation of energy' theorem, Proposition 41, to determine the path of a light ray after it had entered the terrestrial atmosphere.

Early in 1695, however, he revised his calculation to take account of a density gradient which required a more realistic, but much more complicated function of distance from the Earth's centre, involving inverse square and exponential components. He assumed that Boyle's law would hold, with a proportionality between atmospheric density and pressure, and applied the analysis of a spherically symmetric static atmosphere in an inverse square law gravitational field that he had already produced in *Principia*, Book II, Proposition 22, to derive an exponential law of pressure variation which could be reduced by approximation to the form $P = P_0 e^{-z/H}$, where P_0 is the pressure at the Earth's surface, z is the height above the Earth's surface and H is the scale height of the atmosphere.[17]

Since the calculation did not yield a neat analytical solution on integration; the mathematical complexity of the work made Newton withhold anything but the final numerical results from Flamsteed,[18] from whom he had obtained data, though he had succeeded in obtaining a correct result. Lehn writes of Newton's numerical solution: 'When one considers that a 17th century hand-calculated numerical integration is being compared with that of a modern MATLAB program, the agreement is actually quite

---

[14] *Tabula Refractionum*, 1695, Halley (1721), *Corr.* IV, 95.

[15] Biot (1836). His source for Newton's *Correspondence* was Baily (1835).

[16] Theorem 1 in *Of Refraction & ye velocity of light according to ye density of bodies*, MS in private possession, mid-1680s, *MP* VI, 422.

[17] Newton to Flamsteed, 15 January, 26 January, 16 February 1695, 15 March 1695, *Corr.* IV, 67–69, 73–75, 86–88, 93–95.

[18] Lehn (2008, H101).

remarkable.'[19] He concludes that 'Newton fully understood the refraction of the isothermal atmosphere and that he was the first to produce a correct mathematical model for the phenomenon.'[20] From a practical point of view, however, the extra mathematical complexity had been unnecessary, for as Woollard and Clemence write in their textbook on *Spherical Astronomy*: 'to a zenith distance of about 80°, it is almost immaterial what hypothesis is adopted for the rate of increase of atmospheric density, [since] all give practically the same values of refraction.'[21]

Though Newton's method remained unpublished, Brook Taylor revived the application of Proposition 41 to atmospheric light bending in his *Methodus Incrementorum* of 1715,[22] and the rather complex calculation was refined by many subsequent workers; Euler, in particular, made a major contribution by taking account of temperature variation in 1754.[23] Then Laplace, in the tenth book of his *Mécanique Celeste* in 1805, 'raised this 'force' theory of the path of a light-*quantum* to its highest pitch of sophistication'; and, according to Whiteside, 'there is no finer summary of Newton's 'gravitational' theory of the transmission of light.'[24]

The analogy between gravity and refraction has continued to be fruitful up to the present day. Gravitational lenses were discovered in 1979 by the radioastronomers Walsh, Carswell and Weymann, from measurements taken at Jodrell Bank, though they had been postulated earlier in the 20th century.[25] Refsdal and Surdej, in their review article on such lenses, comment that: 'It is interesting to note that gravitational fields in the universe deflect light rays in a way that is very similar to the refraction properties of the lower atmospheric air layers: because of significant temperature and density gradients near the ground, light rays often undergo significant bendings.'[26] It is also equally interesting that the analogy with gravity had been the basis of Newton's original calculation of atmospheric refraction. The concept of gravitational refraction, which was responsible for all of the gravitational effects on light signals discovered in the 20th century, was a concept that originated in the work of Newton and its main results are fully compatible with Newtonian science.

---

[19] *Ibid.*, H 103.
[20] *Ibid.*
[21] Woollard and Clemence (1966, 84–85); quoted Whiteside (1980, 308).
[22] Taylor (1715, 108–118).
[23] Euler (1754, 131ff).
[24] Laplace (1805); Whiteside (1980, 306, 314).
[25] Walsh, Carswell and Weymann (1979).
[26] Refsdal and Surdej (1994, 122).

5.4. Calculation of the Refraction Force

If Newton's expression for the force of refraction at least *implies* a potential energy term of the form mc^2 for light, it is his *quantitative* analysis of the phenomenon which brings him close to the modern understanding of the concept of mass–energy. Query 30 specifically invoked a *transformation* process between light and matter whose origin lay in Newton's realisation of the immense strength of the force involved in refraction, especially when compared with the force of gravity acting on the heavenly bodies. A force of such strength could, he believed, surely effect the transformation which his unitary theory of nature seemed to demand. He was well aware, as we are today, that one of the most important manifestations of this strength was the very great size (in standard units) of terms with c^2 as a factor, and if he did not refer directly to light particles as sources of 'energy', he did consider them as sources of momentum, a term which in relativistic quantum physics is entirely equivalent, and probably more significant.

The calculation of the force per unit mass involved in refraction suggested that it had to be more powerful by many orders of magnitude than the force per unit mass due to the gravitation of the Earth or the force per unit mass due to the gravitation of the Sun.

> And considering how much rays of light are bent at their entrance into pellucid bodies, we may reckon that the attractive power of a ray of light in proportion to its body is as much greater then the gravity of a projectile in proportion to its body as the velocity of the ray of light (squared) is greater then the velocity of the projectile (squared) and the bent of the ray greater then the bent of the line described by the projectile, supposing the inclination of the ray to the refracting surface and that of the projectile to the horizon to be alike.[27] (Here, 'squared' is added to represent the correction '*bis*' made by hand to the Latin version of 1706.)

The earliest version of the calculation appeared in a draft for Query 22 of the Latin *Opticks* of 1706.[28] In the same manuscript draft Newton also asserted that it was the small size of the particles of light which made them particularly active and, indeed, the chief source of activity within matter:

> Now since light is the most active of all bodies known to us, & enters the composition of all natural bodies, why may it not be the chief principle of activity in them? Attraction in bodies of the same kind & vertue ought to be strongest in the smallest particles in proportion to their bulk. 'Tis much stronger in small magnets in proportion to their bulk then in great ones.

---

[27]Draft Q 22 for *Opticks*, 1706; Bechler (1974, 201).
[28]Draft Q 22 for *Opticks*, 1706, Cambridge University Library, Add. MS 3970, f. 292$^\mathrm{v}$, NP.

For the parts of small ones being closer together unite their forces more easily.[29]

The fixed and unchangeable nature of the rays of light, through all the processes to which they were subjected, implied that they were composed of fixed and unchangeable particles of matter; and, if light was made of small bodies, then it must be particularly active on that account, for small bodies are more active than large.

The version later published, which was transferred to the new Query 21 in 1717, further elaborated on this argument:

> As Attraction is stronger in small Magnets than in great ones in proportion to their Bulk, and Gravity is greater in the Surfaces of small Planets than in those of great ones in proportion to their bulk, and small Bodies are agitated much more by electric attraction than great ones; so the smallness of the Rays of Light may contribute very much to the power of the Agent by which they are refracted.

And in a later draft of the refraction calculation, he wrote:

> And therefore since the rays of light are the smallest bodies known to us (for I do not here consider the particles of aether) we may expect to find their attraction very strong. And how strong they are may be gathered by this rule [the centripetal forces are as the squares of the velocities divided by the radii] to be above an hundred million of (millions of) millions of millions of times greater in proportion to the matter in them then the gravity of Earth towards the Sun in proportion to the matter in it. ('Millions' is added here to take account of numerical correction made later.)[30]

The calculation went through a number of iterations and took a number of years to perfect, probably because Newton regarded it as a relatively peripheral activity. Originally, it was based on terrestrial gravity: 'And by this proportion I reckon the attractive force of rays of light above ten hundred thousand (million) millions of times greater then the force whereby bodies gravitate on the surface of this earth in proportion to the matter in them supposing that light comes to us from the Sun to us in about 7 minutes of an hour.'[31] (Again, the substitution of 'million' for 'thousand'

---

[29]Draft Q 21/22 for *Opticks*, 1706; Bechler (1974, 201–202). The role that Newton suggested here for light corpuscles is now assigned to virtual bosons, which are photons in the case of the electromagnetic interaction.

[30]Draft Q 22 for *Opticks*, 1717, Add. MS 3970, f. 621ʳ; Bechler (1974, 207).

[31]Draft Q 22 for *Opticks*, 1706; Bechler (1974, 202).

incorporates numerical correction made later.) Subsequently, it was based on the gravitational attraction of the Earth by the Sun.

Early versions had both arithmetical and conceptual errors but eventually Newton made the necessary corrections, writing in a draft of Query 22/30 that:

> upon a fair computation it will (be) found that the gravity of our earth towards the Sun in proportion to the quantity of its matter is above ten hundred million of millions of millions of millions of times less then the force by wch a ray of light in entering into glass or crystal is drawn or impelled towards the refracting body.... For the velocity of light is to the velocity of Earth in Orbis magnus as 58 days of time (in which) the Earth describes the (same space —); that is an arch equal to the radius of its orb to about 7 minutes, the time in wch light comes from ⊙ [the Sun] to us; that is as about 12,000 to 1. And the radius of the curvity of a ray of light during it(s) refraction at the surface of glass on wch it falls very obliquely, is to the curvity of the earth Orb, as the radius of that Orb to the radius of curvature of the ray or as above 1,000,000,000,000,000,000 to 1. And the force wch bends the ray is to the force wch keeps the earth or any Projectile in its orb or line of Projection in a ratio compounded of the duplicate ratio of the velocities & the ratio of the curvities of the lines of projection.[32]

Newton thus finds that the force per unit mass on a ray (or corpuscle) of light in refraction is above 10^{26} times that of the gravitational force per unit mass exerted by the Sun on the Earth, a figure in agreement with modern values for the respective sizes of the gravitational and electromagnetic interactions. Assuming that the same variation in aether density gradient applies in the two cases, the fact that the Sun–Earth distance is 10^{18} times greater than the range of refractive forces, means that the ratio over the same refractive range must actually be closer to 10^{44}. In another calculation in the same manuscript, Newton takes the radius of the Earth's orbit as 69 million miles (based on a solar parallax of 12 seconds) and the radius of curvature of the path of a light particle as 10^{-6} inches.[33] Assuming that the light from the Sun takes 7.5 minutes to reach the Earth and that in this time the Earth would have travelled 6,197 miles, he finds the ratio of the forces to be about 5×10^{26}. The ratio of 10^{26} or 10^{27} (and 10^{44} over the same range) was an unprecedented figure for an order of magnitude calculation in physics, and became the first of the so-called 'large numbers' representing the ratios of events on astronomical and atomic scales.

---

[32] Draft Q 22 for *Opticks*, 1717; Bechler (1974, 208).
[33] CUL, Add. MS 3970, f. 621^{r}; Bechler (1974, 220).

Though Newton calculated the refractive force per unit mass over a radius of curvature of 10^{-6} inches, his statement that the force ratio was greater than 10^{27} implied that the actual radius of curvature was probably less than 5×10^{-7} inches and, therefore, more than an order of magnitude below the optical wavelengths of 10^{-5} inches which he had used in the early versions of the calculation in Cambridge University Library, Add 3970, f. 292$^\mathrm{v}$, but of the same order as the molecular distances which he had found from experiments on pressing glass slides and lenses together. This suggests that, by the time of the revised calculation at least, he had come to believe that the force of refraction took place at the *molecular* level. Hall states that the figure for molecular diameter which entered into these calculations represents 'the smallest dimension that entered into 17th century theoretical physics.'[34]

Newton's unprecedented calculation of the very great strength of the refracting force was specifically intended to show that it was strong enough to effect the transmutation between light and matter. The argument involving forces also led to the statement, subsequently incorporated into Query 31, that matter was organised at a succession of different levels, with weaker forces at successively higher levels of organisation. His finding that the force per unit mass involved in refraction was 10^{27} times that of the gravitational force per unit mass exerted by the Sun on the Earth contained a hidden value for the mass of the light particle not significantly different from the relativistic masses now attributed to photons. Both the qualitative and quantitative, conclusions which Newton drew from this research have been vindicated by modern developments; their startling similarity to modern ideas is due to the basic validity of the particle concept he adopted.

5.5. The Mass of a Light Corpuscle

The fact that the intermolecular distance was derived directly from the study of optical interference phenomena made it possible to compare the force of capillary action at this distance directly with the optical force of refraction, to which Newton believed it was fundamentally connected. The revised draft of the refraction calculation on f. 621$^\mathrm{r}$ of the Cambridge University Library manuscript Add 3970 is immediately followed by a statement about the capillary forces involved in the Torricellian barometer experiment which seems to have been intended to link the two forces on account of their short range and great strength: 'The Atmosphere by its weight presses the

---

[34]Hall (1981, 157).

Quicksilver into the glass, to the height of 29 or 30 Inches. And some other Agent raises it higher, not by pressing it into the glass, but by making its parts stick to the glass, & to one another.'[35] Experiments by Francis Hauksbee and others had already shown that shaking the mercury in a Torricellian tube caused emission of light at the same time as overcoming the force of cohesion, suggesting a possible link between optical and capillary effects. In the edition of 1717, this passage was incorporated directly into Query 31 almost immediately preceding the description of Hauksbee's experiments.

Now, subsequent to his calculation of the force per unit mass involved in refraction, Newton directed Hauksbee to perform an experiment on capillary action on a drop of oil of oranges which enabled him to make a calculation of the absolute force of cohesion at molecular distances. In this experiment, two glass slides were placed so that they touched at one end, with the upper slide horizontal and the lower one at an inclination. A drop of oil of oranges placed on the lower slide moved towards the concourse of the slides until the force of capillary action on the drop was exactly countered by its weight. The weight of the drop then gave a measure of the force of capillary action. At this time, Newton believed that the forces involved in refraction and cohesion had essentially the same electrical origin, so it is conceivable that he could have combined the two measurements to give a first-order estimate of the mass of a particle of light. Whether or not he did so, calculations give values of order 10^{-35} or 10^{-36} kg, which is close to the relativistic mass of a photon in the violet region of the spectrum (0.29–0.57×10^{-35} kg). A result of this kind obtained long before the quantum theory or high energy physics would have been one of the most remarkable in the whole history of the subject.

The capillary experiment led Newton to calculate that, at a distance of 10^{-7} inches, the force would be sufficient to support a column of the fluid 52,500 inches long.

> ...furthermore the force of attraction is equal to the weight of a cylinder of the oil whose base is the same as the base of the drop and whose height is to 33.44/80 inches as 1/80 is to the thickness of the drop. Let the thickness of the drop be 10^{-7} inches and the height of the cylinder will be $33.40 \times 105/64$ or 52,250 inches, which is 871 paces; and the weight of this will equal the force of attraction.[36]

---

[35] CUL, Add. MS 3970, f. 621^r; reproduced in Bechler (1974, 186).
[36] *De vi Electrica*, c 1712, *Corr.* V, 368.

The thickness of 10^{-7} inches, which was introduced into the account by way of illustration, was specifically chosen as representing the typical distance over which cohesive forces could be expected to act, and it is clear that Newton intended his calculation to represent the value of force at the *molecular* level, or at one in which it made sense to speak of a minimal unit with the required physical properties of the oil. In presenting the experiment in Query 31 Newton first calculated the force at one inch diameter and then added: 'And where it is of a less thickness the Attraction may be proportionately greater, and continue to increase, until the thickness do not exceed that of a single Particle of the Oil.'

The experiment indicates that a drop of oil of diameter 10^{-7} inches will support a cylinder of volume

$$\frac{\pi \times 10^{-14} \times 52250}{4} \text{ cubic inches}$$

or $6.72 \times 10^{-15} \, \text{m}^3$. Taking the density of the oil as $842 \, \text{kg m}^{-3}$, which is that of d-limonene, its major constituent, this is equivalent to a weight of $5.55 \times 10^{-11} \, \text{N}$.

To compare the force of capillary action with the force per unit mass involved in refraction c^2/r, we need to take both forces over approximately the same distance r, and, from Newton's figures, we may suppose this to lie somewhere between 10^{-7} and 5×10^{-7} inches. In a draft Conclusion to the *Principia*, Newton says:

> In this case let the width of the drop of oil be a third of an inch, and its weight will be about a seventh of a grain and its force of attraction toward the glass to the weight of grain will be as 120 to 7 or about 14 to 1. Further, if the thickness of the oil were one million times smaller, and its width a thousand times greater, the force of the total attraction — now increased (in accordance with the aforesaid experiment) in almost approximately the ratio of the square of the diminished thickness — would equal the weight of 140,000,000,000,000 grains; and the force of attraction of the circular part, whose diameter would be a third of an inch, would equal the weight of 14,000,000 grains or about 30,000 ounces, that is, 2,500 pounds. And this force abundantly suffices for the cohesion of the parts of a body.[37]

Newton appears to be suggesting that we take 1/3,000,000 inches as the thickness or diameter of one of the cohering parts of the oil drop, which we can take as either more or less equivalent to our molecules, or as the smallest volume of oil which has the required physical property. The diameter of the

---

[37]Draft Conclusion to the *Principia*, translated in Cohen and Whitman (1999, 287–292).

surface area of the oil film would then be 1000/3 inches. The area of the oil film then becomes $\pi \times 1,000,000/36$ inches^2 and that of one 'part' of the oil $\pi/36,000,000,000,000$ inches^2, implying that there are $10^{18}/1296$ 'parts'. So, the weight of one of the parts will be $2.5 \times 1296 \times 10^{-15} = 3.24 \times 10^{-12}$ pounds weight $= 1.47 \times 10^{-12} \times g = 1.44 \times 10^{-11}$ N. Now, if we take the speed of light c as 69×10^6 miles or 1.11×10^{11} m per 7½ minutes, or $2.47 \times 10^8\,\mathrm{ms}^{-1}$, which is Newton's own value, then c^2/r for $r = 1/3,000,000$ inches is equivalent to $7.19 \times 10^{24}\,\mathrm{N\,kg}^{-1}$. Taking $1.44 \times 10^{-11}/7.19 \times 10^{24}$ gives us a mass of 2×10^{-36} kg. By comparison, the wavelengths of light in the visible spectrum range from about 750 nm to about 390 nm, which is equivalent to a 'relativistic mass' between 2.9×10^{-36} kg to 5.7×10^{-36} kg.

A similar result can be obtained in the following way. For a distance r of 10^{-7} inches, we obtain

$$c^2/r = 2.4 \times 10^{24}\,\mathrm{Nkg}^{-1}$$

for $c = 2.47 \times 10^8\,\mathrm{ms}^{-1}$. Assuming, now, that essentially the same force is responsible for both capillary action and refraction, we obtain

$$\frac{5.55 \times 10^{-11}}{2.4 \times 10^{25}} = 0.23 \times 10^{-35}\,\mathrm{kg}$$

for the mass of a particle of light. For $r = 2 \times 10^{-7}$ inches, the mass becomes 0.91×10^{-35} kg, and for $r = 3 \times 10^{-7}$ inches the mass is 2.1×10^{-35} kg.

Now, the force per unit mass involved in refraction is at least 10^{27} times stronger than the force per unit mass exerted by the Sun on the Earth. For a time period T of 1 year (3.16×10^7 s) and solar distance R, this means that the force per unit mass involved in refraction is

$$\frac{4\pi^2 R}{T^2} \times 10^{27} = \frac{4\pi^2 \times 1.13 \times 10^{11} \times 10^{27}}{(3.16 \times 10^7)^2},$$

assuming Newton's value of 70 million miles for the solar distance, which is 1.13×10^{11} m. This works out to $4.47 \times 10^{24}\,\mathrm{J\,kg}^{-1}$. Assuming that essentially the same force is responsible for both capillary action and refraction gives

$$\frac{5.55 \times 10^{-11}}{4.47 \times 10^{24}} = 1.24 \times 10^{-35}\,\mathrm{kg}$$

as an upper limit on the mass of a particle of light. Using 93 million miles for the solar distance would reduce this to 0.93×10^{-35} kg.

We don't know if Newton ever calculated a mass for light particles from his experimental data. There was certainly, however, just such a calculation in the 18th century, and it was made by Newton's editor, Samuel Horsley, partly using data from Newton himself. In 1767, Horsley argued that: 'The thickness of the particles that reflect the white light of the first order is, according to Sir Isaac Newton, 3 2/5 millionths of an inch.'[38] Newton had estimated that the particles of reflecting metallic surfaces are less than those of glass in the ratio of 2:7, so the thickness of these must be 0.97144 millionths of an inch. For particles of light to enter the pores of metallic substances, they cannot be bigger than 0.145716 millionths of an inch. However, light particles 'positively swarm' into the pores of metals, so they must be much smaller than this, say 10^{-12} inches. Horsley also reasoned that we can stare at a hot iron poker but not at the sun. For a density of sunlight at, say, three times that of the iron, the quantity of matter contained in a particle of light 10^{-12} inches in diameter must be as $1/15552 \times 10^{36}$.

As he explained in 1770, this is because, the 'number of such spherules,' 10^{-12} inches in diameter, that contain as much matter as an iron ball of one yard diameter will be 15552×10^{36}; with the density of each particle (following the argument over the poker) supposed three times less than that of iron. 576×10^{36} 'such spherules contain as much matter as an iron ball of 1 foot diameter.'[39] If the density of iron is taken as $7860 \, \mathrm{kgm^{-3}}$, Horsley's light particles are approximately $1.6 \times 10^{-37} \, \mathrm{kg}$ in mass. This is a fairly good result when compared with modern values in the region 2.9–$5.7 \times 10^{-36} \, \mathrm{kg}$. Some other 18th century calculations produced results that were considerably less 'good'. Benjamin Martin, for example, found a mass $<6.5 \times 10^{-11} \, \mathrm{kg}$, and Pieter van Musschenbroek, a mass $= 2 \times 10^{-13} \, \mathrm{kg}$.

Horsley's calculations were intended as a riposte to Benjamin Franklin, who, in 1752 had written a letter, later read at the Royal Society, questioning the corpuscular theory on the grounds that the Sun had shown no diminution of its dimension over centuries, even though particles were being steadily emitted from its surface, and the planetary orbits had similarly remained unchanged, even though the loss of mass from the Sun would have reduced the attractive force exerted by that body. Franklin had also noted that light appeared to exert no detectable pressure due to a particle momentum.[40] The idea that the Sun would lose mass if light were particulate was actually an old one, and Newton, as an undergraduate, had taken notes on it from the *Physiologiae Peripateticae* (1642) of Johannes Magirus.[41] Horsley's

---

[38] Horsley (1767).

[39] Horsley (1770, 419).

[40] Franklin (1752).

[41] Westfall (1980, 84).

calculations, however, suggested that the Sun was losing only 5×10^{-15} of its mass per year. According to modern estimates, it loses $4 \times 10^9 \, \mathrm{kgs^{-1}}$ or 6×10^{-14} of its total mass, so Horsley's explanation of Franklin's observations is, in principle, correct. Horsley additionally calculated the momentum of each light particle to be $1/151144 \times 10^{27}$ of that which could be imparted to an iron cannon ball, 1 foot in diameter; again, this compares favourably with modern estimates of light particle momenta, at about $10^{-27} \, \mathrm{Ns}$.

5.6. The Interconvertibility of Light and Matter

There is a good argument for saying that the mass–energy relation is inherent within classical Newtonian dynamics itself. Newton's conception of mass incorporated the notion that it also represented a *force* (the *vis inertiae* or inertial force); and force occurs in Newtonian physics in all the contexts in which we now use energy and momentum. Newton's alternative conception of force as impulse makes it virtually identical to momentum in many contexts. In particular, the *vis inertiae* was subject to the law of action and reaction, like other forces, and so, in our terms, was also subject to the conservation of energy. In Newtonian physics, then, the conservation of mass necessarily implies the conservation of force (the third law of motion) because force of any kind is nothing but the action of *vis inertiae* in particular circumstances; and the conservation of (zero) force, which is the essential meaning of Newton's third law, necessarily implies conservation of momentum and of energy.

After the recognition of the equivalence of mass and energy, the next stage is the recognition of their *interconvertibility*. Newton's unitary view of nature, as well as his long-standing work in alchemy, undoubtedly made him prepared to consider this in connection with light and matter. Bodies certainly affect light, bending its rays by gravitational action, and also reflecting, refracting, diffraction, absorbing or emitting them; and light affects bodies, probably being 'the chief principle of activity in them,' since the absorption or emission of light is involved in so many other processes.

> For the matter of things is one and the same, which is transmuted into countless forms by the operations of nature, and more subtle and rare bodies are by fermentation and the processes of growth commonly made thicker and more condensed. By the same motion of fermentation bodies can expel certain particles, which thereupon by their repulsive forces are caused to recede from each other violently; if they are denser, they constitute vapours, exhalations and air; if on the other hand they are very small they are

transformed into light. These last adhere more strongly since bodies do not shine save by a vehement heat.[42]

'... and so great a force in the rays cannot but have a very great effect upon the particles of matter with which they are compounded for causing them to attract one another and for putting them in motion amongst themselves...'[43]

If matter was made of corpuscles, then, surely, so was light. There could not be two fundamentally different types of substance. Though matter and light each had properties which made them seem distinct, the unity between them was manifested by their mutual interactions. 'Certainly there is some spirit hid in all bodies, by means of which light and bodies act upon each other mutually.'[44] Of course, it is usual now to think of 'bodies' as material entities and light as a kind of 'immaterial' form of energy. Bodies, we say, have a 'rest mass' as well as a dynamic momentum, whereas light photons have only the dynamic momentum. There are also people who would like to reserve the term 'mass' only for rest mass and exclude the concept of 'relativistic mass', as used for the total energy, combining rest mass and momentum. Ultimately, however, all mass is dynamic, even the so-called 'rest' masses of particles like electrons and protons, derived as they are from such things as *zitterbewegung* and dynamic gluon exchange. Quantum field theory does not distinguish the particles like electrons and W or Z bosons, which have rest mass, from those without, like photons and gluons. They are all excitations of quantum fields, and the particular dynamics which determine how their total mass–energies are distributed, depend on the particular symmetries applicable to the fields. Newton is, therefore, completely modern in seeing no fundamental distinction between light corpuscles and material particles and in proclaiming that they determine each other's compositions.

The early versions of Query 22/30 stressed the strong connection between the transmutability or interconvertibility of light and matter and the role of light as the 'active power' (in modern terms, the 'energy source') within matter: 'Do not bodies & light mutually change into one another. And may not bodies receive their most active powers from the particles of light wch enter their composition (of wch they are composed *cancelled*)?'[45] A unitary view would also make interconversion a *mutual* process — particles of light not only possess mass but actually make up matter — and from precisely

---

[42] Draft *Conclusio* for *Principia*; *Unp.*, 341–342.

[43] Draft Query 22 for *Opticks*, 1706; McGuire (1968, 160).

[44] *De vi Electrica, Corr.* V, 365.

[45] Early version of Q 22/30 for *Opticks*, CUL Add. MS 3970, f. 292^r; Bechler (1974, 200).

such reasons we now infer that the equation $E = mc^2$ applies not only to light but also to 'solid' matter. In both the draft Query and in the published *Opticks* this became the argument, already quoted, leading to the smallness of the rays of light determining the strength of their activity.[46]

In an earlier *Praefatio* for the *Principia* Newton had argued:

> From the largest particles, then, sensible bodies are formed which allow light to pass through them in all directions, and differ markedly among themselves in density, just as water may be 19 times rarer than gold. The particles of such bodies can very easily be agitated by a vibrating motion and this agitation may last a long time (as the nature of heat requires); through such agitation, if it is slow and continuous, they little by little alter their relative arrangement, and by the force by which they cohere they are more strongly united, as happens in fermentation and the growth of plants... If however that agitation is vehement enough, they will glow through the abundance of the emitted light...[47]

It is anything but a coincidence, therefore, that the draft Query in which Newton produced the calculation of the force per unit mass ratio for the refraction of light was also the one in which he first argued that light could be converted into matter, and matter into light, and further declared that light was the active principle in all organised matter. His calculation of the very great strength of the refracting force was specifically intended to show that it was strong enough to effect the transmutation. According to the historian Zev Bechler, the whole argument led to the statement, now in Query 31, but originally part of this draft, that matter was organised at a succession of different levels, with weaker forces at successively higher levels of organisation;[48] and it stemmed from a mass model of dispersion in which the largest particles were the least accelerated, a kind of partial return to his earlier views on the inverse relationship between wavelength and momentum. Of course, his alchemical studies, reflected in the passage from the draft *Conclusio* to the *Principia*, and also in such manuscripts as *Of Natures Obvious Laws & Processes in Vegetation*, were already leading him in this direction.[49] By splitting up this draft between several Queries and suppressing some of the material altogether, Newton gave the impression that its various conclusions were merely oracular pronouncements on various subjects of interest, prescient to be sure, but not stemming from any unified

---

[46] Q 21.
[47] Draft *Praefatio*; *Unp.*, 306–307.
[48] Bechler (1974, 192ff).
[49] *Of Natures Obvious Laws*, NPA.

theory of light and matter, and bearing only a coincidental resemblance to certain modern developments.

The argument on the hierarchical structure of matter, which found its way into Query 31, originated with this first calculation of refraction. Immediately below the calculation on f. 292^v in Cambridge University Library MS Add 3970, Newton drafted a passage which in the printed version later became:

> Now the smallest Particles of Matter may cohere by the strongest Attractions, and compose bigger Particles of weaker Virtue; and many of these may cohere and compose bigger Particles whose Virtue is still weaker, and so on for divers Successions, until the Progression end in the biggest Particles on which the Operations in Chymistry, and the Colours of natural Bodies depend, and which by cohering compose Bodies of a sensible Magnitude.

Bechler has explained how this hierarchy of particle structures and cohesive interactions could arise 'through a kind of screening effect, in which the bulk of the particles could screen their neighbours' forces on account of their short range.'[50] Newton had already considered such effects in the body of the *Principia*, and an early draft intended for a 'fourth book' of the *Opticks* which was never published indicates that he had calculated that the critical case involved forces inversely proportional to powers of the molecular separation greater than 4, using arguments that are very similar to ones still used today.[51] Bechler's analysis has shown that this led to forces on composite bodies which were nonlinear functions of mass, explaining why the smallest bodies were subject to the strongest forces; and allowed Newton to propose a model of chromatic dispersion based on the mass differences between the particles responsible for light of different colours.

The hierarchy proposal of Query 31 also recalls the passage in the suppressed *Conclusio* to the *Principia*, already quoted, in which 'fermentation' causes bodies to expel particles, which recede from each other violently on account of repulsive forces, and can appear as vapours or gases, but when they are very small they can be transformed into light.[52] We have already seen how Newton effectively calculated the relative energies per unit mass associated with gas molecules and light particles in Query 21. It is, therefore, of particular interest that the first version of this argument, with the specific

---

[50]Bechler (1974, 198).

[51]Draft for *Opticks*, Book IV, CUL Add. MS 3970, f. 336^r; Cohen (1966, 180); Bechler (1974, 192ff).

[52]Draft Q 22 for *Opticks*, 1706; McGuire (1968, 160).

use of the relation

$$\text{velocity}^2 = \frac{\text{elasticity}}{\text{density}}$$

occurs immediately after the revised refraction calculation on f. 622$^\text{v}$ of Add 3,970.

5.7. The Newtonian Aspect of $E = mc^2$

Though the origin of the rest energy term in special relativity has never been fully established, it has long been thought that it is a specific result of that theory, that it cannot be derived from 'classical' approaches, and that it is somehow bound up with the intrinsically 4-vector nature of relativistic space-time. It has often been thought inconceivable that Newton could have had any such idea in mind when he wrote Query 30; and, in any case, a classical calculation of kinetic energy, even for the case of solid light particles of mass m and velocity c, would still yield a value only half of that which mysteriously appears as a result of relativistic calculations. Newton's statement there has been thought to show only a generic similarity to the modern ideas, and to have no link in respect of calculation. But the drafts for Query 22 of the Latin edition of the *Opticks* (1706), which ultimately became Query 30 of the subsequent English edition (1717), are based very strongly on the idea that the refraction of light is due to a force mc^2/r acting over distance r; and it is only a short step from a force of this form, acting under steady-state conditions, to a *potential* energy of the form mc^2.

In fact, $E = mc^2$ emerges, as we have seen, from the special theory of relativity because it is a hidden truth in classical physics, built into the structure of mass and inertia. Its appearance under relativistic conditions is a result of the fact that potential energies are not visible until they are needed, because we normally only observe differences and not absolute values. The manuscripts of the famous series of Queries which Newton added to his *Opticks* between 1704 and 1717 contain calculations which are equivalent in modern terms to the use of the equation $E = mc^2$ for photons in the theory of relativity. These calculations can now be seen to be the basis for Newton's remarkable published statements on the interconvertibility of light and matter. The apparent coincidence between relativistic and classical viewpoints in this and other areas of Newtonian and post-Newtonian physics can be explained in terms of the fundamentally classical origin of the mass–energy relation; and it was a series of precise calculations, rather than merely qualitative arguments, which made Newton believe that the electro-optic force was a possible mechanism for the transition.

The appearance of ratios and terms involving c^2 or mc^2 in these calculations is not only coincidental in its resemblance to modern ideas. As we have seen, Newton, like his immediate successors, worked from an analogy between optics and steady state dynamics in which mc^2 was a *potential*, rather than kinetic, energy term. The analogy produced results meaningful in modern terms because of the validity of the virial theorem, a general dynamical principle known to him and derived by him in the *Principia* (Book I, Proposition 40, Corollary 2). Because of the truth of the virial theorem, Newtonian relations involving a force term such as mv^2/r necessarily presuppose an equivalent energy equation involving terms of the form mv^2.

In this sense, the concepts used by Newton and by modern physicists are exactly the same, even though modern physicists are not aware of their classical origin. If Newton's terminology is largely unfamiliar to us and places emphasis on different aspects of similar material, largely through the use of force as a concept rather than energy, his fundamental line of reasoning is not significantly different to that of modern theory. The current preference for momentum, rather than energy, as the basic concept in relativistic quantum theory even provides a strong point of contact.[53] Newton was always able to develop physical theories that were based on pure abstractions and so could be applied to areas well beyond those with which he was immediately concerned; fundamental physics is only strong when it is able to recognise its own abstract basis.

Those aspects of relativity which show 'surprising' similarities to Newtonian physics are not, therefore, the result of 'coincidence' or 'accident'. They occur because new information has uncovered the hidden structure of classical physics. Relativistic results which coincide with classical ones must necessarily be classical in origin, even if it takes relativistic reasoning to uncover them. Though the equation $E^2 = p^2c^2 + m^2c^2$ is an expression of relativistic or Lorentzian invariance, and parallels the relativistic relation between time, space and proper time, $c^2t^2 = r^2 + c^2\tau^2$, the relationship between the *conservation properties* of mass, momentum and energy is nothing to do with relativity as such, but is rather a hidden assumption of Newtonian physics. And relativity introduces the term mc^2 to make its energy equations compatible with the conservation laws in the Newtonian limit when $c \gg v$. Many seemingly 'relativistic' and 'quantum' results have entirely classical parallels, because they stem from symmetries and dualities built deep into the structure of physics at a fundamental level.

---

[53] In the case of the photon, of course, there is no distinction between these two quantities.

The direct relationship between the total mass of a body and its energy content, proposed by Einstein on inductive grounds, is undoubtedly true, but, as Einstein found out, it cannot be derived from relativity theory. It works because it merges with the classical limit. The work of Newton and some 18th and 19th century scientists, though containing no such explicit relation, tells us why it operates also at a classical level. It also explains the curious fact that photon energy is mc^2, rather than $mc^2/2$, being apparent in the 18th century calculations of gravitational light bending, as well as in Boltzmann's 19th century gas-analogy for radiation pressure.

Newton's optical drafts show a whole host of relationships between apparently unconnected branches of physics and, because of the generality and abstractness of the mathematical formalisms on which they are structured, allow us to make large generalising connections between ideas developed by different scientists at different periods. Significantly, in studying them, we learn that the fundamental meaning of a scientific truth is not necessarily to be found in the context in which it was first discovered. Newton's views on the relation between light and matter, and between force and material structure, are not merely brilliant speculative guesses, but exact consequences of his corpuscular theory of light, backed up with precise mathematical calculations which differ in no significant respect from those based on special relativity and quantum theory. Though there are superficial differences in their presentation as mechanistic models, the Newtonian and relativistic quantum theories of the mass–energy of the photon yield exactly the same values, in principle, for significant quantities because they are both based on the same more fundamental abstract mathematical truths.

Chapter 6

Quantum Theory

6.1. Periodicity as a Quantum Phenomenon

Quantum mechanics has produced a profound change in physics since its introduction by Heisenberg in 1925.[1] It is the only development in physics since the 17th century that can genuinely be called revolutionary. In quantum mechanics as opposed to classical physics, there are no such things as material particles positioned in space acting on each other with various forces. Matter is defined only by probability distributions; forces are the results of interactions with a quantum field; the space and momentum variables constituting the phase space of classical mechanics, and uniquely specifying the behaviour of classical systems, cannot be defined precisely at the same time. As such it would appear that quantum physics is the antithesis of Newtonian thinking which is firmly grounded on the behaviour of material particles interacting with each other from defined positions in space. Newton's own thinking, however, was never purely 'Newtonian' in the way this came to be defined by his successors, and the concepts of quantum mechanics seem surprisingly close to his deepest speculations. Many particular aspects of quantum mechanics were anticipated in his work in one way or another, as he took his reasoning to the utmost limit that could be maintained by the state of contemporary knowledge. Though he was unable to provide a synthesis of these ideas, he was convinced that such a synthesis existed, and made no attempt to bury the discordant facts under a catch-all mechanistic hypothesis. The problems would have to remain until some new general principle or principles emerged. They could not be resolved without one.

Of course, prior to the introduction of quantum mechanics, there was a more restricted quantum *theory*, due to Planck, in which energy was emitted

---

[1] Heisenberg (1925).

or absorbed by a system only in discrete packets or quanta, and this had been extended by Einstein to give a material or particulate character to the quanta, or as they were later called 'photons'.[2] Einstein was fully aware that this related to aspects of Newton's well-known corpuscular theory of light, but few then believed that Newton's light corpuscles had much more than a coincidental resemblance to the new packets of electromagnetic energy. This was because Newton had never produced a systematic account of a corpuscular theory of light, which, however seemingly unavoidable, was still a mechanistic hypothesis, and what passed for a fundamental Newtonian theory of optics had been largely carried out by others.

In 1738 Robert Smith, Plumian Professor of astronomy and experimental philosophy at Cambridge in succession to Roger Cotes, published an influential *Compleat System of Opticks*, which interpreted Newton's corpuscular theory of light in purely mechanistic terms, with light corpuscles acting as material particles subjected to short-range forces.[3] Clairaut produced another mechanistic treatment of light corpuscles in 1741,[4] based on the equivalent force-based refraction equation to Proposition 94 from Book I of Newton's *Principia*:

$$\frac{\sin i}{\sin r} = \sqrt{1 + \frac{k}{v_i^2}}.$$

At the end of the 18th century Laplace added his massive authority to this tradition, finally, against all expectation, deriving a corpuscular theory of double refraction, the optical effect that seemingly only made sense within a wave theory.[5] Newton, however, had never regarded the corpuscular theory as anything more than a probable hypothesis and he knew that there were aspects of light behaviour which did not respond well to mechanistic reasoning. These unresolved aspects would eventually require quantum theory for their full explanation, and it is fascinating to see Newton using his inductive method to take optical physics to the point where only quantum explanations would suffice. Apparently unreconcilable evidence forced him to make extraordinary claims, which happen to be true!

---

[2] Einstein (1905); Lewis (1926) for the term 'photon'.
[3] Smith (1738).
[4] Clairaut (1741).
[5] Laplace (1805); Whiteside (1980, 306, 314).

Newton does not describe light as corpuscular in the body of the *Opticks*. In the Queries, he asks:

> Are not the Rays of Light very small Bodies emitted from shining Substances? For such Bodies will pass through uniform Mediums in right Lines without bending into the Shadow, which is the nature of the Rays of Light. They will also be capable of several Properties, and be able to conserve their Properties unchanged in passing through several Mediums, which is another Condition of the Rays of Light.[6]

In the earlier *An Hypothesis Explaining the Properties of Light*, he states a corpuscular theory of light only as a probable hypothesis.

> ...I suppose light is neither ...aether nor its vibrating motion, but something of a different kind propagated from lucid bodies. They that will ... may suppose it multitudes of unimaginable small and swift corpuscles of various sizes, springing from shining bodies at great distances one after another, but yet without any sensible interval of time, and continually urged forward by a principle of motion.[7]

However, in the main text of the *Opticks*, he uses the term 'rays' for its 'least Parts'. His definition of a ray right at the beginning of the work (Definition 1) is 'quantum' in the modern sense of being the minimal discrete quantity, rather than being specifically particulate in the material sense:

> By the Rays of Light I understand its least Parts, and those as well Successive in the same Lines, as Contemporary in several Lines. For it is manifest that Light consists of parts, both Successive and Contemporary; because in the same place you may stop that which comes presently after; and in the same time you may stop it in any one place, and let it pass in another. For that part of Light which is stopp'd cannot be the same with that which is let pass. The least Light or part of Light, which may be stopp'd alone without the rest of Light, or propagate alone, or do suffer any thing alone, which the rest of Light doth not or suffers not, I call a Ray of Light.[8]

Newton's least parts would appear to be unequivocally a quantum concept, for, even in a wave theory, gauge invariance or the lack of any absolute phase means that individual waves cannot be measured; measurement is only possible when the light is modulated, or the waves are grouped in some way. Newton's insistence on the fact that light, as observed, can only

---

[6] Q 29.

[7] *Hyp., Corr.* I, 370.

[8] Book I, Part 1, Definition 1.

be described if we consider its identifiable parts is therefore perceptive, as well as completely correct. His theory of light rays then becomes effectively a *quantum* theory, and not just another version of the centuries old corpuscular theory, through his insistence that the material nature associated with the rays is due to a colour-producing property, which is implicitly associated with *mass*.

However, because light also has periodic properties, as the Newton's rings experiment unavoidably demonstrates, and these have to operate at the same time as the mass-related conservation of refrangibility, then some kind of probabilistic interpretation appears to be another necessary consequence. The theory of fits is also, therefore, quantum in this second sense, and the fits of easy transmission and easy reflection have been compared to the transition probabilities or probability wave amplitudes of quantum mechanics. Whittaker says of Newton's theory that: 'It was however a remarkable anticipation of the 20th-century quantum-theory explanation: the "fits of easy transmission and easy reflection" correspond to the transition probabilities of the quantum theory.'[9]

Similar statements have been made elsewhere. *Encyclopaedia Britannica* says of Newton that:

> He accepted a concept of the luminiferous ether, and he postulated that the particles had 'fits of easy reflection' and 'fits of easy transmission,' i.e. he assumed that they changed regularly between (1) a state in which they were reflected at a light surface and (2) a state in which they were transmitted. He thus introduced periodicity — one of the basic ideas of wave theory — in a form that anticipates the quantum mechanics.[10]

W. A. B. Evans comments that: 'One cannot help but feel that Newton's early ideas about his light corpuscles possessing 'Fits of easy transmission' and 'fits of easy reflection' (in order to explain interference) is not too distant from the modern concept of the oscillating phase of the probability wave amplitudes that guides our modern corpuscle, the photon.'[11]

The very expressions 'easy transmission' and 'easy reflection' automatically conjure up a probabilistic, rather than deterministic, interpretation, in accord with Newton's own remarkable verbal statement, recorded by David Gregory in May 1694: 'A ray of light has paroxysms of reflexion or refraction,

---

[9] Whittaker (1951–1953, 21–22).
[10] *Encyclopaedia Britannica*, 15th edition, 1986, 23: 2.
[11] Evans (1987, 233).

and indeterminate ones at that.'[12] The astonishing choice of words here shows that Newton was quite explicit about indeterminacy, a concept which also links up with the modern idea of gauge invariance, indeterminacy of the fits being exactly equivalent to the indeterminacy of phase. As in so many other cases, the extraordinary compatibility of Newton's seemingly strange ideas with much more recent ones is the product of his method of working inductively direct from the evidence, and using analytical abstraction, regardless of whether it would fit with any preconceived hypothetical model. Synthetic developments from a mechanistic or hypothetical model are only as good as the model, however, mathematically sophisticated it might be. Analytical results, however, will be limited only by the quality of the experimental data. They may be partial or seemingly incoherent but in the end they will point in the right direction.

6.2. Polarisation and Diffraction

Newton's own analysis of the behaviour of light in reflection and refraction is also readily expressed in quantum terms using such 20th century ideas as the pilot wave: 'To explain how particles can produce interference patterns, Newton invented what Louis de Broglie would later call the pilot wave, which accompanies and even precedes the electron, giving it "fits and starts" of easy and difficult transmission.'[13] Though Newton knew nothing about electrons, he did know about photons, or their equivalent, and frequently discusses the dualistic aspects of light in terms that are strongly suggestive of a pilot wave.

> For assuming the rays of light to be small bodies emitted every way from shining substances, those when they impinge on any refracting or reflecting superficies, must as necessarily excite vibrations in the aether, as stones do in water when thrown into it. And supposing these vibrations to be of several depths or thicknesses, accordingly as they are excited by the said corpuscular rays of various sizes and velocities; of what use they will be for explicating the manner of reflection and refraction, the production of heat by the sunbeams, the emission of light from burning putrifying or other substances whose parts are vehemently agitated, the phenomena of thin transparent plates and bubbles, and of all natural bodies, the manner of vision, and the difference of colours, as also their harmony and discord, I shall leave to their consideration, who may think it worth their endeavour to apply this hypothesis to the solution of phenomena.[14]

---

[12]D. Gregory, *Annotations Mathematical, Physical and Theological from Newton*, 5, 6 and 7 May 1694, *Corr.* III, 339.
[13]Weaver (1987, 1: 511).
[14]Newton to Oldenburg, 11 June 1672 in answer to Hooke, *Corr.* I, 174.

Just as Newton's theory of fits is now seen as an anticipation of the probabilistic features of quantum mechanics, so also is his explanation of polarisation. While fits solved the problem of what we now describe as interference phenomena, an experiment on Iceland spar, which had defied even Huygens' brilliant explanation of double refraction using nonspherical pulses, had to have another explanation. In this experiment, performed in 1672 or 1673, Huygens passed the two rays emerging from a crystal of Iceland spar through a second crystal, arranged with faces parallel to the first, and found that the ordinary ray was again refracted as an ordinary ray, while the extraordinary ray was refracted as an extraordinary ray. But a rotation of the crystal through 90° converted the ordinary ray to an extraordinary ray, and the extraordinary ray to an ordinary one; while intermediate positions, produced both ordinary and extraordinary rays from each of the two incident rays.[15]

Huygens searched for a modification of light produced by the crystal to explain this, but Newton attributed the effect to an original and unchanging property inherent in the rays or particles of light, concluding that light particles must have 'sides' transverse to their direction of motion, and that it was the direction of these sides relative to certain axes in the crystal that determined whether they would be transmitted directly or refracted at an angle. 'Every Ray of Light has therefore two opposite Sides, originally endued with a Property on which the unusual Refraction depends, and the other two opposite Sides not endued with that Property.' In a draft of Query 26, he notices: 'The difference between the two sorts of rays in the Experiment mentioned in the 25th Question, was only in the positions of the sides of the rays to the planes of perpendicular refraction. For one & the same ray is here refracted sometimes after the usual & sometimes after the unusual manner according to the position which its sides have to the crystalls.'[16] There was only one kind of ray, not 'two sorts of Rays differing in their nature from one another,' and it was the position of this ray with respect to the crystal which determined the manner in which it was refracted.[17]

The property was in some sense a 'polar' one: 'And since the Particles of Island-Crystal act all the same way upon the Rays of Light for causing the unusual Refraction, may it not be supposed that in the formation of this Crystal, the Particles not only ranged themselves in rank and file for concreting in regular Figures, but also by some kind of polar Virtue turned

---

[15] Huygens (1912, 92–94).
[16] Draft of Q 26, for *Opticks*, CUL, MS Add. 3970.3, 263ʳ, NP.
[17] Q 26.

their homogeneal Sides the same way.'[18] And, according to Query 29, the peculiar property was like that of the magnetic 'virtue' which lines up the poles of two magnets:

> ...the unusual refraction of Island-Crystal looks very much as if it were perform'd by some kind of attractive virtue lodged in certain Sides both of the Rays, and of the Particles of the Crystal...since the Crystal by this Disposition or Virtue does not act upon the Rays, unless when one of their Sides of unusual Refraction looks towards that Coast, this argues a Virtue or Disposition in those Sides of the Rays, which answers to, and sympathizes with that Virtue or Disposition of the Crystal, as the Poles of two Magnets answer to one another.

Thomas Young claimed that Newton's hypothesis was merely a 'verbal solution',[19] and some modern authorities have been inclined to follow this line of argument, but Query 29 influenced Étienne Malus, a corpuscular theorist, to introduce the term 'polarisation' into optics on the basis of Newton's analogy with magnetic poles.[20] Young was subsequently partly responsible for suggesting that the polarisation was due to the transverse nature of light waves, a fact demonstrated with mathematical sophistication in the wave theory of Fresnel.[21] Longitudinal waves cannot possibly be polarised, and Newton was fully aware of the inability of Huygens' longitudinal wave theory to explain polarisation using this model.

Newton, of course, had no idea that the polarisation was a property of the transverse nature of light waves, or that this transverse component could, in some way, be represented as magnetic. The suggestion of a polarising process, however, is still a brilliant intuition, despite Young's comments and those of the many commentators who have followed his line of argument. Newton does not say that the polarising virtue *is* magnetic. He goes as far as he can with the evidence in presenting an abstract analysis, while leaving his options open to await future developments, and stating that his current theories do not depend on the assumption of such facts as hypotheses. The 'virtue' might or might not be magnetic, but it is at least similar to a magnetic one. As on other occasions, he denies that his theory implies a certain possibility while giving the impression that he may be prepared to consider it. He leaves a lead which a successor can follow, when more evidence becomes available,

---

[18] Q 31.

[19] Young (1817b); *Works*, 1: 325.

[20] Malus (1809).

[21] Young (1817a, 1817b); Fresnel (1816, 1821); Buchwald (1989, 205–206); Wood (1954, 202).

instead of creating a hypothesis which can only be discarded when it fails to square with the facts.

In Derek Gjertsen's judgement, as opposed to that of Young, Newton's assumption of the strange 'polar' properties of the light rays within a crystal of Iceland spar led him to 'account much better than Huygens for the facts of double refraction.'

> Looking back with hindsight after Malus [Gjertsen says], Newton's extraordinary physical intuition is revealed yet again. What he wrote in Queries 25-6 was sketchy and speculative and, for a century, led nowhere. Despite this, it contained the essential insight into the nature of double refraction, and provided the basic physical understanding which would permit the explanation of much more besides.[22]

Newton was both astonishingly insightful and prophetic when he wrote at the end of Query 25: 'And it remains to be enquired, whether the Rays have not more original Properties than are yet discover'd.'

Having come across the power of conservation principles, and correctly privileging them over other approaches to physics, Newton was obliged to follow the logic of his own reasoning even when the facts appeared to be anomalous. The 'conservation' property or lack of modification on refraction meant that the 'rays' of light had to have some of the physical characteristics that we generally associate with particles; and the new 'polar' property required of these 'corpuscles' would eventually be identified as an intrinsic angular momentum or spin.

Of course, a successful explanation of polarisation in terms of classical waves does not preclude a more fundamental one in terms of quantum theory, and a modern authority has put Newton's explanation into this context:

> By studying the flight of photons through polarizers of Iceland spar Newton discovered the internal structure of the photon that we call polarization. His photon has sides transverse to its direction of flight, like a playing card moving parallel to itself. When the 'long' side of the photon, for example, is parallel to a certain grain in the Iceland spar, the photon passes through. When its long side is perpendicular to the grain, the photon is refracted and its path bent across the grain.[23]

This can even be related again to the idea of the pilot wave: 'Newton anticipated the paradigm of quantum physics when he asked what happens in the intermediate cases where the long side of the photon is neither precisely

---

[22]Gjertsen (1986, 181).
[23]Weaver (1987, 1: 511).

parallel nor perpendicular to the grain,' for in that case the transmission or refraction of individual corpuscles cannot be predicted. 'Once again the pilot wave intervenes to make fits and starts of easy and difficult transmission; the results cannot be predicted. Indeed, the same quantum principles are involved in both kinds of beam splitting.'[24] The 'polar' explanation becomes yet another example of Newton's inductive method of analytic abstraction, verified by subsequent application of Ockham's razor. It is *qualitative*, rather than merely verbal. As in so many other cases, the result may look strange in the context of 17th-century mechanistic science, but it is entirely compatible with the findings obtained via non-mechanistic modern approaches.

In addition, it would seem from at least one manuscript, already quoted, that transverseness of a kind might be associated with the subsidiary vibrations produced by the rays of light. Though this did not lead at the time to an explanation of polarisation in terms of wave optics, it does now suggest another quantum connection. This is in Newton's explanation of the phenomenon of *inflection*, now explained as diffraction, but regarded by Newton as a force produced at the 'sharp edges of dense bodies,' producing 'an undulating motion like that of an Eele.'[25]

A reduced version of this survived in Query 3 of the *Opticks*: 'Are not the Rays of Light in passing by the edges and sides of Bodies, bent several times backwards and forwards, with a motion like that of an Eel? And do not the ... Fringes of colour'd Light above-mention'd arise from ... such bendings?' On this, Sir Michael Berry has written:

> One way of writing wave equations, discovered in the context of quantum mechanics by Madelung and emphasized by Bohm and his followers, is in terms of the local current vector rather than the function describing the wavefield. The lines of current can be regarded as analogous to the rays of geometrical optics, but survive into wave optics. Where propagating waves interfere, these rays indeed wriggle like an eel, as the result of non-Newtonian forces acting from edges etc. Although (perhaps for reasons of historical contingency) this is not the interpretation of wave physics that most of us use, all wave phenomena can be regarded as the effect of these generalized rays. So, Newton was right after all![26]

Here, the rays are interpreted as the streamlines of the wavefunction or the lines of current density, and these, as Berry has shown, display Newton's eel-like undulations. Even stronger forces would bend the streamlines into

---

[24] Weaver (1987, 1: 511).
[25] Manuscript connected with *Opticks*, 1690s, Propositions 12–14; Westfall (1971, 412).
[26] Berry (1997).

closed orbits or optical vortices, which are phase singularities. These require a reflecting screen for observation, rather than a black, opaque one. The streamlines are normals to a type of wavefront introduced by Young in the form of cotidal lines; these are 'contours of constant phase of the total wave' rather than lines drawn normal to the rays, as in the more usual type of wavefront from geometrical optics. For wave equations, written in terms of current density, the 'streamlines are influenced non-locally by a 'quantum potential' in addition to the Newtonian force.'[27]

The broadening of the diffraction pattern observed in Newton's single-slit diffraction experiments also has a quantum explanation in Heisenberg's uncertainty principle.[28] According to the Heisenberg principle, the product of the uncertainties in momentum (Δp) and position (Δx) of a photon cannot be reduced to a value below about the value of Planck's constant h (in more accurate terms, $\hbar/2$). So, the two physical quantities describing classical phase space cannot be defined with unlimited accuracy at the same time. The uncertainty in position for a photon passing through a slit of width d is given by $\Delta x = d$, while if θ_1 is the angle of the first minimum in the diffraction pattern, and mc the momentum of the photon, the uncertainty in momentum is given by $mc \sin \theta_1$, which, using the de Broglie relation between momentum and wavelength, becomes $(h/\lambda) \sin \theta_1$. Now, the angle of the first minimum can be found from $\sin \theta_1 = \lambda/d$. So, in principle, $\Delta p \Delta x$ is of order h. Reducing the slit width, and so reducing Δx, will make the pattern spread out and so increase Δp. Increasing the slit width will have the reverse effect.

6.3. Aethereal Effects and Quantum Phenomena

In yet another striking approach to quantum ideas, Observation 8 of *Opticks*, Part II includes a suggestion of what is now known as frustrated total internal reflection or the Goos–Hänchen shift, an optical analogue of quantum mechanical tunnelling.[29] If light is sent at the angle for total internal reflection through two prisms placed close together but separated by a small gap of air (or some other substance which is less refracting than glass), then some of the light that should be reflected back by the first prism, will 'tunnel' across the gap and go out through the second prism. As the incident photons spread into the gap, we see the light beam being displaced laterally along the

---

[27] Berry (2002).

[28] Heisenberg (1927).

[29] *Opticks*, Part II, Observation 8; see also Observation 1.

surface of the first prism after penetrating the gap, with a shift in the point of reflection. The lateral shift postulated by Newton was the first indication that the Law of Reflection was not precisely correct in quantum terms, and was measured for the first time by Goos and Hänchen in 1947.[30]

Newton's suggestion seems to be closely related to his views on reflection and refraction, and even 'inflection' (or diffraction), as involving the entire surface and not being due to particle 'contact'. In a draft conclusion to the *Principia*, which again brings in the 'eel-like' motion due to inflection, he writes:

> We have shown in the *Opticks* that light is not reflected and refracted in only a single point, by falling upon the thick and solid particles of bodies, but is curved little by little, by a spirit that lies hidden in the bodies. Rays of light passing through glass and falling upon the further surface of the glass are partly reflected and partly refracted, even if the glass is placed in a vacuum. If the glass is immersed into some oil which abounds with the greatest force of reflecting, such as is oil of turpentine or oil of flaxseed or of cinnamon or of sassafras, the light which in a vacuum would be reflected passes out of the glass into the oil, and in nearly straight lines — reflection and refraction ceasing — and therefore is not reflected and refracted by the parts of the glass, but passes directly through the whole glass; and (if the oil is absent) passes into the vacuum before it is reflected or refracted, and therefore is reflected or refracted in a vacuum through the attraction of the glass, that is, through the action of some spirit that lies hidden in the glass and goes out of the glass into the vacuum to some small distance and draws back the rays of light into the glass. Further, that the rays of light are inflected in the vicinity of bodies, and at some distance from the bodies and without any contact with the bodies at all, is most certain from phenomena. They are inflected indeed with a serpentine motion, now approaching the bodies, now receding from them, and this is so whether the bodies are pellucid or opaque. And hence it is concluded also that the rays are agitated by some tremulous spirit in the vicinity of bodies. That this spirit moreover is of an electric kind is obvious from what has been said above.[31]

It is significant that in Newton's physics, as in physics today, there are two types of force explanation on offer, the local forces between particles of matter, and the non-local forces supplied by an aether ('vacuum' in modern parlance), usually in the form of a density gradient. The most spectacular example of the second is Newton's aethereal explanation of cohesion, which is astonishingly similar in qualitative terms to those of modern quantum theorists. Though this originally used the Cartesian-type gravitational aether

---

[30] Goos and Hänchen (1947).
[31] Draft Conclusion to the *Principia*, translated in Cohen and Whitman (1999, 290–291).

that he had abandoned before he wrote the *Principia*, the reasoning would apply equally well to the electrical aether that he adopted after writing the *Opticks*.

An Hypothesis Explaining the Properties of Light has the explanation:

> So I suppose æther, though it pervades the pores of chrystal, glass, water, and other Naturall bodyes, yet it stands at a greater degree of rarity in those pores then in the free æthereall Spaces, & at so much a greater degree of rarity as the pores of the body are Smaller.... This also may be the principall cause of the cohæsion of the parts of Solids & Fluids, of the Springines of Glass & other bodyes whose parts Slide not one upon another in bending, and of the Standing of the Mercury in the Torricellian Experiment, sometimes to the top of the Glass, though a much greater height then 29 inches. For the Denser æther, wch Surrounds these Bodies, must croud & presse their parts together much after the manner that Air surrounding two Marbles presses them together if there be little or no Air between them.[32]

Similar ideas are put forward in the letter to Boyle of 28 February 1679:

> When two bodies moving towards one another come neare together I suppose ye æther between them to grow rarer then before, & ye spaces of its graduated rarity to extend further from ye superficies of ye bodies towards one another, & this by reason yt ye æther cannot move & play up & down so freely in ye strait passage between ye bodies as it could before they came so neare together.[33]

Qualitatively, these suggestions bear a remarkable similarity to the modern explanation of the van der Waals force, and, in particular, to the well-observed Casimir effect, based on the existence of the zero-point energy in the vacuum.[34] Here, we regard each point in empty space as supplying the equivalent of $\frac{1}{2}h\nu$ of zero-point energy for every conceivable vibration frequency ν, where h is the quantum of action introduced by Planck. In the confines of a restricted space, say between two molecules or between two metal plates, the longer wavelength, lower frequency, modes of vibration are excluded, so the 'free' spaces beyond have a higher energy density. The radiation pressure differential thus produced forces the molecules or plates towards each other with a force that depends on the inverse fourth power of the separation. We may note here that an inverse fourth power dependence

---

[32] *Hyp., Corr.* I, 367.
[33] Letter to Boyle, 28 February 1679, *Corr.* 1: 366–367 and 2: 289–290.
[34] Casimir (1948).

on the separation is the minimum requirement for a force of cohesion that Newton derived separately on inductive grounds at a later period.

Two phenomena that are observed in classical contexts but are equally essential to quantum physics were also considered by Newton. These are radiation pressure and radiation reaction, which he understood in connection with the probably corpuscular nature of light. Newton was aware that Kepler had attributed the movement of comets away from the Sun as being produced by rays of light from the Sun 'carrying along with them the matter of the comets tails.'[35] Early on, in *Quaestiones Quaedam Philosophicae*, he proposed a question on light pressure, with the 'rays' now definitely implying corpuscles: 'Whither ye rays of light may not move a body as wind doth a mill saile.'[36] Mill sails were something he would have known about from his childhood, and his comment may remind us of the Crookes radiometer, where the idea is put into effect, through a subtle connection between radiation pressure and heating. He also discussed Kepler's theory in his own work on comets.

Radiation pressure additionally connects with radiation reaction, due to the emission of light, and to the heating effect that results on the emitting body. The draft conclusion for *Principia*, describing in detail the actions of an electric spirit (seemingly a kind of emanation or radiation), referred to mysteriously at the end of the General Scholium which was added in the second edition, shows a clear anticipation of this concept.

> This spirit, therefore, if sufficiently agitated, emits light, and so emits light in extremely hot bodies and in turn suffers a reaction from the emitted light. For whatever acts on something else suffers a reaction on itself. This is confirmed by the fact that all bodies grow warm in the light of the sun, and when sufficiently warmed emit light in turn. And certain phosphors are aroused in the light of the sun (or even in the light of clouds) to emit light, and that without the heat of a thick body. Between bodies and light there are action and reaction through the mediation of the electric spirit.[37]

A form of radiation reaction is also invoked in a draft Query: 'Pellucid substances act upon the rays of light at a distance in refracting reflecting & inflecting them, & the rays mutually agitate the parts of those substances at a distance for heating them, & this action & reaction very much resembles an attractive force.'[38]

---

[35] *Principia* I, Proposition 41.

[36] QQP, '*Of light*', f. 103^v, NP.

[37] Draft Conclusion to the *Principia*, translated in Cohen and Whitman (1999, 290).

[38] Draft Q 20, CUL, f. 289^r, NP.

Radiation pressure and radiation reaction connect aether and quantum theories. For example, in 'On the Question of Absolute Velocity and on the Mechanical Functions of an Æther, with some Remarks on the Pressure of Radiation' (1898), Oliver Lodge, a prominent aether theorist of the late-19th century, wrote that: 'it may be found that the property which enables aether to alternately receive and deliver electrokinetic energy is essentially a quasi-material or potentially material property.' Light pressure, in Lodge's view, was possibly 'going to constitute a most important exception, the first of a large class of forces for which the usual interpretation of the third law of motion may have to be enlarged.' In addition, 'a boundary between pulsating and inert aether behaves in some respects like a material partition; it is able on certain conditions to take the place of a reacting body in Newton's third law. This is suggestive in connexion with the view that regards all matter as a variety of aetherial strain and motion.'[39]

Such quantum dynamical phenomena as radiation pressure and radiation reaction also connect with spectroscopy, where Newton recognised the dynamical origin of all spectra, with fixed bodies, heated beyond a certain degree, emitting light and shining, with the emission being performed by the vibrating motions of their parts. However, though the light-emitting motion was sometimes associated with heat, it also occurred in cold bodies: seawater in a raging storm; mercury agitated in *vacuo*; the back of a cat; putrefying wood, flesh and fish; *Ignes Fatui*, or vapours from putrefying waters, stacks of moist hay or corn growing hot by fermentation; glowworms and the eyes of some animals by vital motions; phosphorus agitated by the attrition of any body, or by the 'acid' particles of the air; amber and some diamonds by striking, pressing or rubbing them; etc.[40]

Another connection with Newton's pioneering work on spectra has been found by Duarte in modern quantum optics.[41] Newton, includes in his *Opticks* a diagram showing beam expansion using a single prism; though there is no description of the effect in the text, the angle of incidence in the diagram clearly exceeds the angle of emergence, indicating that the beam has been expanded, and that Newton had discovered the principle around 1670.[42] Another diagram in the *Opticks*, this time accompanied by a verbal description, shows an arrangement of 'optical architecture', with an array of prisms used to add or subtract dispersions and control the direction of the

---

[39] Lodge (1898).
[40] Q 8.
[41] Duarte (2000).
[42] Book I, Part 2, Figure 3 for Experiment 3.

rays.[43] Both beam expansion and multiple-prism dispersion are now used in light tunable lasers. The first high-performance tunable laser was developed by Hänsch in 1972, with a narrow line-width created by beam expansion in a single prism[44]; multiple-prism beam expanders subsequently improved the performance. Duarte comments on these developments: 'The fact that principles outlined nearly three hundred years ago are embodied in a device of the quantum era serves as a dramatic reminder of the profound vision displayed by Newton in his writings on optics.'[45]

6.4. Planck's Constant

There is an even more striking illustration of how close Newton came at times to modern quantum conceptions, driven, as always, by the desire to reconcile apparently irreconcilable facts. If we go back to his earliest work on 'coloured circles', we find an extraordinary statement, already quoted: 'That if ye medium twixt ye glasses bee changed ye bignes of ye circles are also changed. Namely to an eye held perpendicularly over them, the difference of their areas (or ye thicknesses of ye interjected medium belonging to each circle) are reciprocally as ye subtlity of ye interjected medium or as ye motions of ye rays in that medium.'[46] The 'thicknesses of ye interjected medium belonging to each circle' is Newton's 'interval of fits' or wavelength in his earliest work. The 'motions' of the rays are of course the momenta of the corpuscles or photons. Here, momentarily, is an equivalent of our inverse relation between momentum and wavelength, expressed in the equation $p = h/\lambda$, with h the quantum of action. The document in which this occurs is a mass of contradictions, a working document not a finished research paper, and the idea is never referred to again. It flashes into existence like a virtual particle, only to disappear again almost immediately, never to be resurrected. Yet, its momentary existence is remarkable in itself. Of course, the key element in the modern conception of this relation is the application to matter waves, or waves associated with material particles, but, even if only applied to light, this result is the one required to make both the wave model and the particle model give *entirely correct* explanations of the process of refraction, and to reconcile the principles of least time and least action.

A modern interpretation of the Newton's rings experiment would say that, for an air film of thickness t and light of wavelength λ at an angle r to

---

[43] Book I, Part 2, Figure 7 for Experiment 10.

[44] Hänsch (1972).

[45] Duarte (2000, 28).

[46] *Of ye Coloured Circles*; Westfall (1965, 191).

the vertical, dark rings will be observed when

$$2t \cos r = m\lambda$$

and light rings when

$$2t \cos r = (m + \tfrac{1}{2})\lambda,$$

where m represents the integer sequence $0, 1, 2, 3, \ldots$. In de Broglie's quantum interpretation, the momentum and wavelength of a light photon are related by

$$p = \frac{h}{\lambda},$$

where h is Planck's constant.[47] So, the thickness of the film required to produce a given ring for any incident orientation of a light ray will be inversely proportional to the momentum associated with the photons in the ray. At the same time, the geometry of the apparatus in the Newton's rings experiment ensures that the thickness of the film producing any ring is proportional to the area of the ring or to its diameter squared. That is,

$$t \propto D^2 \propto \frac{1}{p \cos r}.$$

In his own accounts of the experiment, Newton is correct, in these terms, in insisting that the size of the rings for light of different colours incident at various angles has an inverse relation with the perpendicular components of the momentum of the deflected light particles, and his experimental data showed a clear inverse proportionality between the area of the ring, and hence film thickness, and the vertical component of momentum.

In the case of obliquity, he uses diameter, rather than area, for the proportionality relation; for different refracting media, however, he uses the inverse proportionality of area or thickness (i.e. wavelength) and 'motion' (i.e. momentum).[48] And, in this very early optical work, he is using something like a conventional wave interpretation, inspired by the pulse theory of Robert Hooke. For a refracting medium, the wavelength is reduced in the medium by the factor n, where n (the refractive index) is the ratio of the velocity of light in vacuum (c) to the wave or 'phase' velocity in the medium (v). As the wavelength is reduced, so the momentum, which

---

[47] de Broglie (1923a, 1923b, 1923c, 1924).
[48] *Of ye Coloured Circles*, Proposition 4, Westfall (1965, 191).

is the ratio of energy to phase velocity, is increased. Assuming constant frequency ν,

$$p = \frac{h}{\lambda} = \frac{h\nu n}{c} = \frac{h\nu}{v}$$

and

$$E = pv.$$

The equation $p = h/\lambda$ thus becomes meaningful in the context of Newton's early work on interference rings, with its emphasis on explanation in terms of light particle momenta, though the opportunity was seemingly lost in the confusion of this early work, and never resurrected. It would occur again in classical optics in the early 19th century controversy between Poisson and Fresnel, but once more without recognition of its fundamental consequence.[49] However, it is plausible to argue that the quantum theory would not seem nearly so revolutionary if it had first arrived in optics or atomic theory rather than the more obscure field of black body radiation.

If we indulge in a little counterfactual history, it is possible to derive a value of Planck's constant from Newton's data of the right order, say within the range 10^{-34}–10^{-33} Js. We could, for example, use the ratios given in *Opticks*, Book II, Part IV, Observation 8 to determine the 'intervals of fits' (say δ) for red and violet rays from the value of $1/89{,}000$ inch for those at the border of yellow and orange given in Book II, Part I, Observation 6. From these, we calculate the Newtonian δ (our $\lambda/2$) as lying between 2.0×10^{-7} m for violet and 3.2×10^{-7} m for red. Then, with 10^{-35} kg for mass and 2.47×10^8 ms for c, we obtain values of $mc\delta$ (our $h/2$) between 4.9×10^{-34} and 7.9×10^{-34} Js. Of course, counterfactual arguments are not serious history. While Newton *could* have derived a value for the mass of a light particle within the context of his possessing values for force and force per unit mass which could have easily been related, and, for a considerable number of years in the latter part of his life, he had momentum increasing for different colour-producing rays while the interval of fits decreased, the derivation of a constant relating momentum and wavelength or equivalent would seem to require a highly improbable sequence of events from both chronological and logical points of view. The real point to be made is that the quantity we now know as Planck's constant was always within the reach

---

[49] Poisson claimed that, in the case of a simple spherical wave, Fresnel's integrals required particle speed to be inversely proportional to wavelength — a possibility that Fresnel regarded as a perfectly reasonable consequence of his theory. Poisson (1823a, 1823b, 1823c, 1823d); Fresnel (1823). A full account of the conflict is given in Buchwald (1989, 188–198).

of classical Newtonian physics and was derivable without needing an input from some more exotic branch of the subject.

6.5. Quantum Gravity

Quantum gravity is considered a very modern concept, an extension of quantum field theory in which all other fundamental forces between particles (electric, strong and weak) are quantum interactions, mediated by virtual particles, described as exchange or gauge bosons. For the electric interaction, the exchange particle is the photon, for the strong interaction there are eight gluons, and for the weak interaction the W^+, W^- and Z^0 intermediate bosons. If gravity is a quantum force of the same kind, then it will require a mediator of the same kind, a 'graviton'. For interactions which are repulsive between identical particles, like the electric, strong and weak interactions, the rules of quantum field theory determine that their mediating bosons are spin 1. However, for an interaction which is *attractive* for identical particles, like gravity, the exchange boson has to be a spin 2 object, a new kind of particle not yet discovered in nature. The complications introduced by this spin 2 property have also ensured that no satisfactory theory of quantum gravity has yet been discovered.

The underlying reason for the difficulty is relatively simple and was actually first pointed out by Newton in a draft of the 1690s. Essentially finite values for physical quantities quickly blow up to become infinite ones, with no apparent means of escape from the difficulty. Newton's draft effectively rejects both quantum gravity and a nonlinear gravitational field:

> If anyone should . . . admit some matter with no gravity by which the gravity of perceptible matter may be explained; it is necessary for him to assert two kinds of solid particles which cannot be transmuted into one another: the one of denser which are heavy in proportion to the quantity of matter, and out of which all matter with gravity and consequently the whole perceptible world is compounded, and the other of less dense particles which have to be the cause of the denser ones but themselves have no gravity, lest their gravity might have to be explained by a third kind and that so on to infinity. But these have to be very much less dense so as by their action to shake apart and mutually scatter the denser ones: by which means all bodies composed of the denser ones would be quickly dissolved. And since the action of the less dense upon the denser will have been proportional to the surfaces of the denser, while gravity arises from that action and is in proportion to the matter of which the denser ones consist, it is necessary that the surfaces of the denser ones must be in proportion to their solid content, and therefore that all these particles must be equally dense and that they can neither be broken nor worn away nor in any manner destroyed;

or else the ratio of the surfaces to the solid content, and consequently the ratio of gravity to the quantity of matter would be changed. Therefore one must altogether determine that the denser particles cannot be changed into the less dense ones, and thereupon that there are two kinds of particles, and that these cannot pass into one another.[50]

Newton realised relatively early on that gravity was different in kind from the other forces that he knew, electricity and magnetism, and the other forces that he suspected must exist. A draft conclusion to the *Principia* states succinctly views that he seems to have held for a considerable time. Gravity only attracts; there is no gravitational repulsion. It cannot be shielded, or increased or diminished in any way, and is not affected by friction. It is universal to all matter and proportional to the quantity of matter or mass, and it acts at long distances. It is not affected, it would seem, by any other agency, such as heat or other forces. Electricity, on the other hand, can be both attractive and repulsive. It can be shielded by ordinary matter, can be increased and diminished, and is greatly excited by friction. It is not universal in the sense that mass is and is not proportional to mass, though it is present in most bodies, ready to be drawn out by friction or other means, such as heating. It generally acts at small distances, though it can be emitted to a large distance. Magnetism, like electricity, can be both attractive and repulsive. It can be shielded by hot bodies, can be increased and diminished in strength, and is more readily transferred by friction. Like electricity it is not proportional to mass. In Newton's day (as is evident from various statements he made) it was only observed in iron, though he also suggested that it might be hidden more generally in bodies in the same way as electricity. Again, like electricity, it only acts at small distances, though 'small' must be at least several feet, and scientists had known since Gilbert's *De Magnete* of 1600 that the Earth itself acted like a magnet.

Gravity was, it would seem, a special force — no other force was universal or directly linked to mass. All other forces could be varied locally, cancelled out by having both attractive and repulsive qualities, shielded, and confined within bodies. Any other force in nature would have to be similar to the known electricity and magnetism because it is not observed universally and only exists in a hidden form. However, because gravity is universal and defined by the mass of objects, it is inconceivable that any kind of material particle could exist without it. So, if the gravitational forces is transmitted by material particles, then these must have gravity. Then

---

[50]Draft Proposition 6, Corollary 3, for *Principia*, early 1690s; McGuire (1967, 72–73).

the gravitational force between these intermediate particles and between these particles and the original gravitating masses must produce yet more intermediate particles and so on to infinity. Quantum gravity has exactly the same problem; it produces 'unrenormalizable infinities', which cannot be removed because, uniquely, the sources are universal to all matter and never negative. Electricity and magnetism are not universal and the exchange bosons or photons are not sources of the electric and magnetic fields. Even where quantum field theory allows a virtual exchange photon to split up into paired particles and antiparticles, which *are* sources of the fields, the opposite charges introduce a degree of cancellation; and, where the strong or 'colour' force requires bosons which are direct sources of the field (the so-called 'coloured gluons'), there is cancellation because of the fact that the overall property (the 'colour') has to sum to zero.

The problem of quantum gravity is not the creation of infinities but the fact that it seems impossible to cancel them by the usual procedure of renormalisation. Renormalisation allows the electric and other forces to produce finite values for the masses and charges of the interacting particles because there is overall cancellation on a large scale between positive and negative charges and other sources, and between forces of attraction and repulsion, so the quantities can be renormalised to finite values, and the force is not cumulative. It is for the same reason that the electric and other non-gravitational forces are normally hidden within matter and can be shielded. Newton's relatively simple but profound argument immediately gets to the core of the problem and appears extraordinarily modern in its context. Though Newton could not identify a cause for gravity, he seemingly sensed its special nature, and the relentlessly cumulative way it builds up to be a large-scale force, while the range of the other forces is reduced by the mutual cancellation of their attractive and repulsive aspects. He would soon produce his calculation showing the relative magnitudes of the forces of gravitation and the inner forces acting on the particles of micromatter, which would show that gravity's large-scale effectiveness was not due to its intrinsic strength, but to the property of relentless accumulation overcoming its weakness.

Renormalisation is frequently seen as a kind of somewhat desperate *ad hoc* process, which cannot be regarded as the true way that nature operates, but which currently provides an interim solution by the cancellation of infinite quantities of opposite sign. Newton has a sense that the universality of gravity means that infinite forces will exist and is more than happy to cancel infinities, as he tells Bentley in a letter of 17 January 1693. Not all infinites, he says, are equal; finite forces may still remain when we subtract unequal but oppositely acting infinite forces from one another. The argument has a generic similarity to the one used in renormalisation, and this, of course, is not a coincidence, because the two arguments use a similar rationale. The cancellation, as he is aware, is only possible because forces are vector quantities which can be opposite in effect, but, as he is also aware, the argument cannot be applied to infinite gravitational masses, or in our terms energies, for these are positive scalars and will simply add up without cancellation.

> But you argue in ye next paragraph of your letter that every particle of matter in an infinite space has an infinite quantity of matter on all sides & by consequence an infinite attraction every way & therefore must rest *in equilibrio* because all infinites are equal. Yet you suspect a parallogism in this argument, & I conceive ye parallogism lies in ye position that all infinites are equal. The generality of mankind consider infinites no other ways then definitely, & in this sense they say all infinites are equal, though they would speak more truly if they should say that they are neither equal nor unequal, nor have any certain difference or proportion one to another.... The falseness of ye conclusion shews an error in ye premises, & ye error lies in ye position that all infinites are equal. There is therefore another way of considering infinites used by Mathematicians, & that is under certain definite restrictions & limitations whereby infinites are determined to have certain differences or proportions to one another.... And so a Mathematician will tell you that if a body stood *in equilibrio* between any two equal and contrary attracting infinite forces, & if to either of those forces you add any new finite attracting force: that new force how little so ever will destroy ye equilibrium & put ye body into ye same motion into which it would put it were these two contrary equal forces but finite or even none at all: so that in this case two equal infinites by ye addition of a finite to either of them become unequal in our ways of recconning. And after these ways we must reccon if from ye consideration of infinites we would always draw true conclusions.[51]

Ultimately, the Newtonian theory of gravity offers something that quantum theory needs but that quantum gravity would necessarily lack: a means of instant communication, a correlation process faster than the speed of light. In Newtonian gravity, the transmission of information is instantaneous, though it occurs through negative, rather than positive energy. A vacuum process might be expected to require negative energy, as in the filled quantum vacuum associated with the Dirac sea of negative energy states. Gravity as a vacuum process would suggest 'gravity/gauge theory correspondence', or duality with the local, matter-based processes

---

[51] *Corr.* III, 239.

associated with the other three interactions, in which gravity appears to provide the same effect as the totality of the others. Contrary to what is often proclaimed in standard textbooks, Newtonian gravity has never been 'replaced' by relativity theory, as the latter is physically meaningless without it, and, surprising as it may at first seem, Newtonian gravity may yet provide us with clues to understanding some of the most puzzling aspects of quantum theory.

6.6. Quantum Mechanics and Newtonian Physics

Quantum theory can mean a number of different things which are connected, but distinct, and it is not always obvious which is the defining characteristic. It can mean that energy is transferred only in discrete packets or *quanta*, but it can also mean that physical processes can only be described in probabilistic terms. Wave-particle duality may be considered an essential component, now applied to distinctly material particles as well as photons, and also some kind of fundamental uncertainty. In relation to the first meaning, there is no abrupt point of transition between Newton's corpuscles of light and the photons of the modern quantum theory, just as there is no really abrupt transition between classical and quantum physics.[52] There may be an accumulation of detail, but the principal characteristics are broadly unchanged, for Newton's 'corpuscular' theory of light was necessarily less mechanistic and so closer to the modern quantum theory than is commonly realised. In developing what later came to be known as the photon theory, Einstein referred to the 'Newtonian emission theory' of light, and spoke as if he were reviving it as an alternative viewpoint to the classical wave theory.[53] Of course, the important extension of wave-particle duality to 'material' particles came only with the theory of de Broglie, and what we call 'quantum mechanics' is mostly connected with material particles, rather than with 'massless' photons. However, the Newtonian corpuscles clearly involve probabilistic aspects and a form of wave-particle duality, and Newton's 'field' approach to optics, in which refraction involves the whole refracting surface

---

[52]In reality, there is no specific body of physics that we can describe as 'truly' classical. Classical physics is simply a series of accommodations for the purposes of observation and simplified description to the single body of physics that is ultimately quantum. It is very often possible to find classical analogues for results that, in the first instance, seem to be purely quantum. To take just one example, it is possible, using a discrete differentiation process based on commutators rather than derivatives and a 'wavefunction' that has an amplitude but no phase factor, to provide a Dirac-type equation for a free fermion that can be interpreted either classically or in quantum terms (Rowlands, 2007).

[53]Einstein (1909).

and not just the particles in immediate 'contact' with the light corpuscle, is suggestive of the kind of reasoning which ultimately led to the Feynman path integral.

A great deal of the behaviour of photons can, in fact, be dealt with using classical mechanics, and mechanical versions of optics have been developed which give perfectly good answers to many fundamental questions. Tangherlini, for instance, has derived the Fresnel reflection and transmission coefficients for normal and oblique incidence, using a particle method, but without direct use of Planck's constant, and so indicated why both the quantum Schrödinger equation and classical Maxwell's equations generate the same coefficients.[54] Evans and Rosanquist have derived from Fermat's principle an optical analogue of Newton's force equation $F = ma$, and derived Snell's law assuming no horizontal variation in the refractive index and a vertical variation that is parallel to free fall in mechanics.[55] That the attractive force between light and medium required by the Newtonian theory of refraction actually exists was shown, for unpolarised light entering a glass prism, by Poynting and Barlow as early as 1904, and, for a laser beam at the surface of a liquid, by Ashkin and Dziedzic in 1973.[56] Poynting and Barlow also used the quartz-fibre torsion balance introduced by C. V. Boys to demonstrate the existence of a tangential stress for a light beam reflected obliquely from a partially absorbing surface, in addition to Lebedev's pressure for radiation normal to the surface. At a later date, they also observed the recoil forces produced on a material body by the emission of heat radiation from its surface.[57]

In addition, at least one aspect of Newton's corpuscular theory of light remains fundamental to the way in which we view quantum theory today. Newton believed that the fixed and unchangeable nature of the rays of light, which was maintained through all the processes, such as reflection and refraction, to which they were subjected, must *necessarily* imply that they were composed of fixed and unchangeable particles of matter. This idea became an intrinsic part of the description of the interaction of light quanta and matter when quantum theory was introduced in the 20th century. Remarkably, it also became the basis on which the two theories were used to explain the same experimental evidence, the line spectra of gases. In principle, the absorption and emission spectra of gases show that a

---

[54] Tangherlini (1975).

[55] Evans and Rosenquist (1986).

[56] Poynting (1905a); Ashkin and Dziedzic (1973).

[57] Poynting and Barlow (1909, 1910).

particulate nature of some kind is necessary, together with the appropriate conservation principles.

In 1832 Brewster, the last supporter of the old corpuscular theory with any significant influence, announced to the British Association that the absorption spectrum of nitrogen dioxide was not continuous but incorporated more than a thousand identifiable dark absorption lines, and supported this with an argument that continuous waves would never be able to explain a discontinuous spectrum.[58] Despite many attempts, line spectra were never satisfactorily accommodated to continuous physics, as Maxwell recognised in 1873 when he stated, again at the British Association, using a metaphor which is remarkable in the context of Newton's own esoteric work on the Egyptian royal cubit: 'every molecule... throughout the universe bears impressed upon it the stamp of a metric system as distinctly as does the metre of the Archives of Paris, or the double royal cubit of the Temple of Karnac.'[59] The oscillators within atoms which were presumed to be responsible for the absorption and emission of light seemed to operate only at fixed frequencies and not over the continuous range which the wave theory would have predicted.

Eventually, after Planck and Einstein had developed the modern quantum theory to explain the laws of black body radiation and the photoelectric effect, Bohr applied it to Rutherford's new model of the atom which had negatively charged electrons orbiting a positively charged nucleus.[60] Rutherford's model, it seemed, would be unstable according to classical electromagnetic theory because continuous waves of energy would be emitted or radiated until the atom collapsed with the electrons spiralling into the nucleus (an interesting analogue to Newton's treatment of inverse square law orbits under resistance); but this problem would not arise if energy could be emitted or absorbed by electrons only in fixed 'particulate' amounts. The discussion of the difficulties of classical theory in explaining line spectra would certainly have formed part of the background to Bohr's development, and to that of Nicholson and other precursors who may have influenced him in applying Planck's quantum theory to the atom.

Bohr's theory of course, had a spectacular success in explaining the emission spectrum of hydrogen. In principle, both the Newtonian 'corpuscular' theory of light and the new, quantum version were used to explain the same

---

[58] Brewster (1832).
[59] Maxwell (1873).
[60] Rutherford (1911a, 1911b); Bohr (1913); Rosenfeld (1963); Rutherford (1913); Pais (1986, 212).

evidence, the line spectra of gases, and ultimately for the same reason. Bohr's theory required the orbitals of electrons to be fixed because the frequencies and energies of the quanta involved in transitions between them were fixed, and could not be replaced by an equivalent amount of energy, supplied continuously. Bohr's light quanta had conserved or immutable properties, in the same way as Newton's light rays had to be particulate because refraction and other processes showed that their properties were also fixed and immutable.

It is frequently made out that modern physics represents a fundamentally different worldview from that of Newton. Newton, we are told, was an instigator of the Age of Reason who believed in a 'clockwork universe'; in the 20th century, however, physicists learned through quantum mechanics that physics is probabilistic and non-deterministic and, by implication, non-Newtonian. But quantum mechanics is not non-Newtonian. It follows all the laws of Newtonian mechanics — the laws of motion, the conservation of momentum and angular momentum, and the conservation of energy. It allows force-type representations of systems like harmonic oscillators which look exactly like the classical ones if the classical quantities are replaced by expectation values, and quantum mechanical particles are still subject to inverse square laws of gravitational and electrical attraction. Even the laws for strong and weak nuclear interactions — forces similar to those which Newton believed to exist deep within matter — are really only modified versions of the same principle with an inverse square and another component, and it may be that, at grand unification energies, only the inverse square component survives. And, if the 4-vector or 'relativistic' nature of the theory makes matter and energy interconvertible, then so did Newton himself.

Also, from the moment Newton began to analyse the three-body problem in gravitation, a non-deterministic element appeared in physics. No longer was it possible, even in principle, to determine the exact positions of physical objects acting under forces. Newton certainly believed that all matter was in motion, stating that particles would only be at rest if they were also absolutely cold. In *Principia*, Book I, Section XI, he wrote: 'I have hitherto been treating of the attractions of bodies towards an immovable centre; though very probably there is no such thing existent in nature.'

It is an interesting reflection, therefore, that there could, indeed, hardly be a more perfect illustration of Newton's principle of conformability in nature than de Broglie's original prediction of matter waves from which quantum mechanics subsequently developed. If waves of light have particle-like properties, and if nature is simple and conformable, then particles of matter should have wavelike properties. Also, it is likely that the abstract non-determinism of wave mechanics would have concerned him much less than it has many 20th and 21st century practitioners of the subject. His gravity theory, in being concerned with an infinite number of instantaneous interactions, is fundamentally indeterminate (as well as nonlocal). He might well have argued: 'If God does not play dice, then we are God.'

The fact is that quantum physics is abstract in precisely the way that Newton's was. Gauge theories and fundamental symmetries would not be incompatible with the abstract approach he adopted towards attractions and repulsions. Renormalisation would not have been alien to him, as his letter to Bentley shows. The problems that quantum mechanics has generated have been entirely due to its intrinsically abstract nature. People have worried themselves over its 'physical' (by which they mean 'concrete') interpretation. Schrödinger's cat and the Einstein–Podolsky–Rosen paradox are results of their authors' inability to come to terms with the fact that gauge invariance does not have a concrete explanation, such as hidden variables or many universes.[61] Quantum theory is in the same position today as gravity was in Newton's, or electromagnetic field theory in Maxwell's — a purely abstract theory which is completely successful in explaining physical effects, but which remains unacceptable to those who remain rooted in the idea that nature is fundamentally concrete.

The interesting thing about the history of modern physics is the fact that it seems to be gradually fulfilling the Newtonian programme in spite of its own manifest preferences for something different. Though Newtonian and other related ideas, such as the Maxwellian theory of the electromagnetic field, were long resisted on account of their intrinsically abstract nature, quantum mechanics has forced modern physicists into the same abstract positions in an amazing repetition of the Newtonian development. When Werner Heisenberg introduced his new mechanics, strongly influenced by the formalised dynamical tradition dating back to Lagrange, in which relations were expressed only between observable quantities, he abandoned the reality of Bohr's physical electron orbits and the concept of orbital radius, in order to retain the measurable quantity of frequency as a fundamental quantity.[62] This led to Bohr's Copenhagen interpretation, in which the abstract system was effectively separated from the physical measuring apparatus.[63] The

---

[61]Schrödinger (1935); Einstein, Podolsky and Rosen (1935).
[62]Heisenberg (1925).
[63]Bohr (1928).

subsequent development of the ideas of non-locality and entangled states, backed up by strong experimental evidence, has led physics back to the indeterminate infinity of interacting states postulated in the Newtonian theory, but ignored by his successors. It begins to look as if modern physics is intent on restoring all of Newton's most extreme positions.

Chapter 7

The Electric Force

7.1. The Electric Force and Aethereal Currents

The most notable product of Newton's inductive method is his theory of the electric force. Here is a wide ranging but almost entirely qualitative theory of the actions of the electric force derived from only the most basic experimental knowledge of electrostatics, which is nevertheless astonishingly accurate and which reaches far into the future. It is significantly a product of his later years — a mature reflection on many decades of thinking on a set of key experimental phenomena which suddenly emerges as a coherent synthesis on the basis of spectacular new experimental data. Surprisingly, though many of the relevant documents have been published and commented upon, most historians have been more interested in their relation to Newton's seeming reversion to an aether theory in the period 1703–1717 than to the evidence they provide for his possession of an uncommon but strikingly effective scientific method in qualitative thinking.

Newton's approach, in his investigation of electricity, to the explanation of all the familiar processes of nature, apart from gravity, by the action of an electric 'spirit', trapped inside bodies, but released by a variety of means as a kind of radiation or emanation, is the culmination of his investigation of the structure of matter and the forces which create its various states. The electric theory is certainly the most underestimated part of his work, though it provides us with a classic case of his analytical method, a significant reason for this being that, though it is the subject of many draft treatises and notes, it hardly crept into print in his lifetime or for 300 years afterwards.

It was obvious to Newton that the fundamental principles that he had derived for explaining celestial phenomena through the long-range gravitational force must apply to a host of other phenomena governed by short-range forces.

> Have not the smallest Particles of Bodies certain Powers, Virtues, or Forces, by which they act at a distance, not only upon the Rays of Light for reflecting, refracting, and inflecting them, but also upon one another for producing a great Part of the Phænomena of Nature? For it's well known, that Bodies act one upon another by the Attractions of Gravity, Magnetism, and Electricity; and these Instances show the Tenor and Course of Nature, and make it not improbable that there may be more attractive Powers than these. For Nature is very consonant and conformable to her self.[1]

From the beginning he took a special interest in the electrical force, which seemed to be more universal than the magnetic:

> Just as the system of the Sun, planets and comets is put in motion by the forces of gravity and its parts persist in their motions, so the smaller systems of bodies also seem to be set in motion by other forces, and their particles to be variously moved in relation to each other, and especially by the electric force. For the particles of bodies seem to be set in motion by the electric force which they possess and to act upon each other at small distances even without being rubbed, and those which are most electric, when rubbed emit a spirit to great distances, by which straws and light objects are now attracted, now repelled and now moved in various ways.[2]

Newton had himself investigated the electrical force in some early experiments. In *An Hypothesis Explaining the Properties of Light*, he writes:

> At least the electric effluvia seem to instruct us, that there is something of an æthereall Nature condens'd in bodies. I have sometimes laid upon a table a round peice of Glasse about 2 inches broad Sett in a brass ring, so that the glass might be about 1/8 or 1/6 of an inch from the table, & the Air between them inclosed on all sides by the ring, after the manner as if I had whelmed a little Sive upon the Table. And then rubbing a pretty while the Glass briskly with some rough and rakeing stuffe, till some very little Fragments of very thin paper, laid on the Table under the glasse, began to be attracted and move nimbly to & fro: after I had done rubbing the Glass, the papers would continue a pretty while in various motions, sometimes leaping up to the Glass & resting there a while, then leaping downe & resting there, then leaping up & perhaps downe & up againe, & this sometimes in lines seeming perpendicular to the Table, Sometimes in oblique ones, Sometimes also they would leap up in one Arch & downe in

---

[1]Q 31.
[2]Draft for General Scholium (MS A), Add 3965, ff. 357–358; *Unp.*, 349–352, translation 352–335.

another, divers times together, without Sensible resting between; Somtimes Skip in a bow from one part of the Glasse to another without touching the table, & Sometimes hang by a corner & turn often about nimbly as if they had been carried about in the midst of a whirlwind, & be otherwise variously moved, every paper with a divers motion. And upon Sliding my finger on the upper Side of the Glasse, though neither the glass, nor inclosed Air below, were moved thereby, yet would the papers, as they hung under the glasse, receive some new motion, inclining this way or that way accordingly as I moved my finger.[3]

The most significant aspect of this is that the attraction of the paper took place on the opposite side of the lens to the one that he had rubbed, showing that the electric attraction acted *through* the glass.

His explanation already has the character of some kind of 'emanation' or 'subtle matter' released from the glass:

Now whence all these irregular motions should spring I cannot imagine, unless from some kind of subtill matter lyeing condens'd in the glass, & rarefied by rubbing as water is rarified into Vapour by heat, & in that rarefaction diffused through the Space round the glasse to a great distance, & made to move & circulate variously & accordingly to actuate the papers, till it returne into the glasse againe & be recondensed there.[4] And...this condensed matter by rarefaction into an æthereall wind (for by its easy penetrating & circulating through Glass I esteeme it æthereall) may cause these odd motions, and by condensing again may cause electricall attraction with its returning to the glass to succeed in place of what is there continually recondensed...[5]

According to Westfall, this early explanation of 'electrical attraction in terms of aethereal currents set up when friction vaporises the aether condensed in electric bodies' actually 'exercised some influence on the future history of electricity.'[6] Turnbull, the first editor of his correspondence, thought that Newton's early experiments, described in the *Hypothesis on Light* of 1675, should have led him to the discovery of the two types of electricity.[7] They appear not to have done so, but his later direct comparison of attraction and repulsion to the algebra of positive and negative quantities was a very important idea which eventually bore fruit in Franklin's concept of positive and negative electricities. 'As in Algebra, where affirmative

---

[3] *Hyp., Corr.* I, 364–365.

[4] *Hyp., Corr.* I, 365.

[5] *Hyp., Corr.* I, 365.

[6] Westfall (1971, 364).

[7] Turnbull, *Corr.* I, 387.

Quantities vanish and cease, there negative ones begin; so in Mechanicks, where Attraction ceases, there a repulsive Virtue ought to succeed.'[8]

The *Hypothesis* was, in some ways, a more publicly acceptable version of a more overtly alchemical manuscript of the 1670s, *Of Natures Obvious Laws & Processes in Vegetation*, where he had written of the concept which he later associated with the source of electricity:

> This is the subtle spirit... this is nature's universal agent, her secret fire, the only ferment and principle of all vegetation. The material soul of all matter which being constantly inspired from above pervades and concretes with it into one form and then if incited by a gentle heat actuates and enlivens it... Note that 'tis more probable the aether is but a vehicle to some more active spirit. ...This spirit perhaps is the body of light because both have a prodigious active principle both are perpetual workers. ...No substance so indifferently, subtly and swiftly pervades all things as light, and no spirit searches bodies so subtly, piercingly, and quickly as the vegetable spirit.[9]

7.2.   A Unified View of the Electric Force

The development of the concept of the vibrating electric spirit to explain the non-gravitational phenomena of nature illustrates the way in which Newton gradually synthesised ideas from an immense variety of sources into an abstract theory of vast potential, one which is recognisably valid even in modern terms; and it may even be possible to identify the precise moment when the synthesis began to take place. The theory emerged from four separate pieces of information, all clearly stated in works of the 1670s:

(1) Light is periodic; this was an unavoidable consequence of the early experiments on thin films.
(2) Air is produced by breaking up particles of matter, and something more rarefied even than air (say, aether) may be produced by breaking up particles even further. This was a result of Boyle's experiments with the air pump extended with the kind of speculation on aethers typical of the time.
(3) Friction produces some sort of electric effluvium from bodies and so must remove something already condensed in them, in the same way as heat produces airs and vapours from solid bodies. This was a mechanistic interpretation of Newton's own experiments on electrostatic attraction.

---

[8] Q 31.
[9] *Of Natures Obvious Laws*, NPA.

(4) A universal agent is the source of 'vegetation', that is the most fundamental form of generation and corruption in both animate and inanimate objects; it is quick to act, subtle, searching and a perpetual worker. Aether, whatever that may be, is only the vehicle for this more active spirit; light has the required properties. This idea was already the result of a synthesis derived from a combination of non-scientific sources, principally alchemy and the Neoplatonic tradition represented by Henry More; the application to light, seemingly made in one of Newton's characteristic moments of insight or 'phase transition' in the brain, was probably suggested by his early concentration on optics.

All these ideas were expressed in a form unique to Newton. Whatever traditional elements they had originally contained, they had all been subjected to processes of transformation and synthesis, and already, in works like the *Hypothesis of Light* of 1675 and the *Letter to Boyle* of 1679, he was attempting to combine them, though still mechanistically, into a unified view of nature. But the attempts were premature and were abandoned after Newton began his radical revision of dynamics in the 1680s with its strong emphasis on a mathematically exact treatment of the concept of force.

Nothing was heard about 'spirits' or 'aethers' in Newton's scientific work for about 25 years, though he continued his investigations in alchemy during this period, and it is difficult to believe that he did not seek to integrate these two areas of study as he had done before. The situation changed in 1703 when Newton became President of the Royal Society, and the Society acquired an experimentalist of genius in Francis Hauksbee. Newton, through his patronage and influence on Hauksbee, may be said to have founded the continuous tradition of research in electricity which began with Hauksbee's experiments.

Whether or not Newton suggested Hauksbee's programme for experiment, the latter certainly used his outstanding technical skill to produce a whole new range of data on subjects which had long been of interest to Newton; and this data gave a completely new meaning to aspects of Newton's earlier work which had been expressed in terms of a universal aether in the *Hypothesis of Light*. However, the point when the potentialities of Newton's earlier work were realised came in 1705–1706 when Hauksbee's experiments showed the startling optical effects produced by his frictional electrical machine. Hauksbee's first appearance at the Society was on 15 December 1703, at the first meeting presided over by Newton. He showed the light produced when mercury was allowed to fall over an inverted glass vessel in the receiver of his air-pump.[10] Although we recognise this now as an example of frictional or triboelectricity, there is no evidence that Hauksbee yet thought in this way. In subsequent experiments, Hauksbee established that the light was a result of the friction rather than intrinsic to the mercury.

Hauksbee then sought to produce even more spectacular effects for the benefit of the Society Fellows. He was not, it would seem, consciously working with electricity when he produced a strong light by using machinery to turn an evacuated glass globe against his hand in a darkened room. Hauksbee thought that he had forced out the particles of light that Newton believed to be part of the composition of solid bodies. Newton, however, recalled his experiments of 1675 and equated the light with the electric effluvia that had seemingly emerged from the underside of his glass lens. The glass globe became a specifically electric generator, like Otto von Guericke's sulphur globe reported in 1672, and Hauksbee's experimental work became channelled into an investigation of the nature of the electric force that had been neglected at the Royal Society for the last 30 years.[11]

Already impressed by Hauksbee's initial experiments, Newton returned to electrical phenomena in the Queries of the Latin edition of the *Opticks* of 1706:

> So also a Globe of Glass about 8 or 10 Inches in diameter, being put into a Frame where it may be swiftly turn'd round its Axis, will in turning shine where it rubs against the palm of ones Hand apply'd to it: And if at the same time a piece of white Paper or white Cloth, or the end of ones Finger be held at the distance of about a quarter of an Inch or half an Inch from that part of the Glass where it is most in motion, the electrick Vapour which is excited by the friction of the Glass against the Hand, will by dashing against the white Paper, Cloth or Finger, be put into such an agitation as to emit Light, and make the white Paper, Cloth or Finger, appear lucid like a Glow-worm; and in rushing out of the Glass will sometimes push against the finger so as to be felt. And the same things have been found by rubbing a long and large Cylinder or Glass or Amber with a Paper held in ones hand, and continuing the friction till the Glass grew warm.[12]

On 6 November 1706, after a demonstration of one of Hauksbee's experiments before the Royal Society, the *Society's Journal Book* records:

---

[10] *Royal Society Journal Book*, IV, 1702–1714, 37; Hauksbee (1709/1719, 9); Guerlac (1972).
[11] Heilbron (1982, 169). Hauksbee's glass globe electric generator has sometimes been attributed to Newton, but it is clearly a product of Hauksbee's mechanical skill in producing friction at a higher rate to produce a more spectacular optical demonstration. The fact that it generated *electricity*, however, seems to be an insight due to Newton.
[12] Added to Q 8 in this edition, English version, published third edition, 1717.

'The president said he thought these Experiments Evinced that Light proceeded from the Subtle Effluvia of the Glass and not from the Grosse body.'[13] The emission of light must be due to the electric force; it was electricity, the 'Effluvia' from the glass, that was fundamental to the process.

Armed with these insights and those outlined in the *Hypothesis* of 1675, Newton was ready to advance an explanation of all the familiar processes of nature, apart from gravity, by the action of an agent which has a strong resemblance to our own electromagnetic field. Passages written 40 years apart, and in totally different contexts, merge almost imperceptibly and show the consistency of his approach, even as his understanding of the nature of this universal agent became ever more precise. This followed from his earlier studies derived from alchemy, into a 'subtil spirit' which is 'Natures universall agent', and the early intuition, which never left him, that this 'subtil spirit' might be the 'body of light'.[14] The spirit was 'diffused' through the 'whole body' of a substance like glass, which was capable of electrification, and surrounded it with a 'small atmosphere' close to the surface which could only be released by friction, but, when released it could act *through* the glass virtually unimpeded.[15]

When Newton had fixed on electricity as the source of light, and, in retrospect, it seems all along that he would eventually come to this conclusion (considering the clear hint of a connection as early as 1675), he then began to reduce his assumptions. Reflection, refraction and diffraction are caused by the same principle as emission, and so they must be electric. Electricity must be in all bodies since all emit light; but not all bodies are electrified by friction, so the electric force must somehow be latent or screened, acting only at small distances. If the forces fall off rapidly with the distance, then they must be very strong at small distances, like the cohesive forces which hold bodies together. The strength is confirmed by experiments in which electrified bodies stick to each other and so cohesion is due to electric forces. Chemical effects are produced by attractive forces of the same kind and so they too must be electric. Also, as Hauksbee had shown, electrical forces may be repulsive as well as attractive just like the forces between the particles of bodies. Magnetism has the same kind of properties and seems to result from the magnetic nature of particles; possibly this also results from the same kind of cause. Finally, the electric force, when liberated by friction, travels long distances, and so must be vibratory; light is vibratory because

---

[13] *Royal Society Journal Book*, X, 140. Westfall (1980, 745).
[14] *Of Natures Obvious Laws*, NPA.
[15] Draft Conclusion to the *Principia*, translated Cohen and Whitman (1999, 289).

it is periodic, and so it must be electric vibrations which emit light. Heat is also vibratory and is present in all substances, so electric vibrations must cause the vibrations of molecules of which heat consists.

Effectively, then, Newton had unified all the known non-gravitational forces into a single electric force, and all the operations of the force were reduced to the action of a single vibrating electric spirit. And this spirit was even tangible, being felt by its action on the hand in Hauksbee's experiments. The whole chain of reasoning can be followed in detail as presented by Newton himself, and it followed entirely from the discovery made in 1705 or 1706 that the production of light was related to the generation of electricity, a discovery which seems not to have been made by Hauksbee himself who, as we have seen, was apparently unaware of the electrical nature of his early experiments.

The decisive break had come with Newton's application of the idea of electric effluvia and the rarefaction of electrical bodies by friction to produce an electrical 'aether' or 'spirit' which had a tangible presence. The disparate elements of the 1670s outlined in the *Hypothesis of Light* suddenly sprang into a unity which required no mechanistic hypothesis. The linking of light with electricity, and the consequent extension of the electrical force to the explanation of all other non-gravitational phenomena in the first two decades of the 18th century, had been prepared for by the experiments and speculations made 40 years before, but it required the trigger of the experiments of 1705–1706, and probably the visual stimulus of actually seeing them performed.

In this way, Newton discovered by analytical reasoning and his recursive method the special role of electricity in the micro-world. All bodies, he believed, have electric properties, however, slight. Drafts began to emphasise the significance of the electrical force. '...gross bodies contain within them a subtle aether or aetherial elastic spirit which by friction they can emit to a considerable distance, and which being emitted is found sufficiently subtle to penetrate and pass through the body of glass, and sufficiently active to emit light at a distance from the gross body if it be there put into a trembling agitation...'[16]

All manner of optical effects are produced by this process:

> This spirit is not terminated in the external parts of bodies at a mathe-
> matical surface, but becomes rarer gradually, and the more rare part of
> it spreads out from bodies on all sides to short distances and gradually

---

[16]Draft Observation XII/III for *Opticks*; Guerlac (1967, 48).

comes to an end. By the action of this spirit the rays are inflected at short distances from bodies, and to a greater extent the closer they pass to bodies; and in permeating transparent bodies they are not reflected and refracted only at a single point but are gradually curved around... And furthermore by acting on this medium they excite vibrations in it in such a way that the bodies, being set in motion, excite vibrations in the air. These vibrations are swifter than light itself and successively pursue and overtake the waves, and in pursuing them they dispose the rays into the alternate fits of being easily reflected or easily transmitted; and when excited by the light of the Sun they excite the particles of bodies and by this excitation heat the bodies; and being reflected at the surfaces of bodies they turn back within the bodies and persist a very long time. But when the hot body is immersed in a cold one they pass out of the hot body into the cold one and very quickly communicate the heat to it. When the rays fall upon the bottom of the eye they excite vibrations in this medium which, carried through the solid capillaments of the optic nerve into the brain, create vision. And just as the vibrations excited in the air by resonant bodies produce all the tones of sound according to their magnitude and number, so the vibrations of the spirit hidden within the capillaments of the optic nerve create all the colours of light according to their magnitude and number. The most refrangible rays produce the shorter vibrations and, in the same time, more of them; these arouse the colour violet; the least refrangible rays produce larger vibrations, fewer in number, which arouse a red colour. The remainder produce intermediate vibrations according to their degree of refrangibility which, in proportion to their magnitude and number arouse the intermediate colours orange, yellow, green, blue and indigo.[17]

The electrically produced light (including that from a gaseous discharge) is a spectacular manifestation of the electrical spirit:

This spirit is...the cause of electrical attraction and not only reflects, refracts and inflects light but also emits it; as is manifest from the following experiment. A glass globe, seven or eight inches in diameter, is exhausted of air and rotated very rapidly upon an axle; if the hand, kept stationary, is rubbed against it, it emits the electric spirit which will shine from the place where the fingers are into the interior of the glass. And if the glass is not wholly void of air the electric spirit issuing forth into the interior of the glass will mingle with the air and will shine like some thin flame filling the whole interior cavity of the glass. And if some white body is placed near the exterior of the glass, near the middle between its poles, separated by about the third or fourth part of an inch from the glass, the electric spirit excited by friction and issuing forth from the revolving glass into the external air and impinging upon that white body will shine, and cause that

---

[17] *De vi Electrica, Corr.* V, 365–366.

white body to glow like a live coal or a piece of putrefying wood in the dark. Now surely the electric spirit which, after it has been forced out of bodies by rubbing them emits light by its vibrations, may be the vehicle by which all flames and all bodies are ignited, and by which putrefying wood emits light.[18]

Newton's early discussions of electrical forces were concerned only with forces of attraction; forces of repulsion were known only in the case of magnetism. But, when Hauksbee's experiments brought cases of electrical repulsion to light, there was then nothing to prevent Newton from identifying the repulsive forces, which he believed to affect matter alongside the attractive forces of cohesion, with forces of electrical origin, and this would have confirmed his supposition that the cohesive forces were electrical also. A draft for the General Scholium of the second edition of the *Principia* listed a number of propositions concerning the forces and the electric spirit responsible for them:

Proposition 1. That very small particles of bodies, whether contiguous or at very small distances, attract one another...

Proposition 2. ...That attraction is of the electric kind.

Proposition 3. That attraction of particles at very small distances is exceedingly strong... and suffices for the cohesion of bodies.

Proposition 4. That attraction without friction extends only to small distances and at greater distances particles repel one another...

Proposition 5. That the electric spirit is a most subtle medium and very easily permeates solid bodies...

Proposition 6. That the electric spirit is a medium most active and emits light...

Proposition 7. That the electric spirit is set in motion by light and this is a vibratory motion, and of this motion heat exists...

Proposition 8. That light falling on the bottom of the eye excites vibrations which, propagated to the brain through the solid fibres of the optic nerves excite vision.

Scholium. That all sensation and all animal motion are accomplished by means of the electric spirit.

Proposition 9. That the vibrations of the electric spirit are swifter than light itself.

Proposition 10. That light is emitted, refracted, reflected and inflected by the electric spirit.

---

[18] *De vi Electrica, Corr.* V, 366.

Proposition 11. That homogeneous bodies are brought together by electric attraction, and heterogeneous ones disassociated.[19]

The General Scholium, as published in the second edition of *Principia* in 1713, added words on this 'spirit' which were for a long time regarded as extremely mysterious:

> And now we might add something concerning a certain most subtle spirit which pervades and lies hid in all gross bodies; by the force and action of which spirit the particles of bodies attract one another at near distances, and cohere, if contiguous; and electric bodies operate to greater distances, as well repelling as attracting the neighboring corpuscles; and light is emitted, reflected, refracted, inflected, and heats bodies; and all sensation is excited, and the members of animal bodies move at the command of the will, namely, by the vibrations of this spirit, mutually propagated along the solid filaments of the nerves, from the outward organs of sense to the brain, and from the brain into the muscles. But these are things that cannot be explained in few words, nor are we furnished with that sufficiency of experiments which is required to an accurate determination and demonstration of the laws by which this electric and elastic spirit operates.[20]

(Motte's English translation of 1729 added the words 'electric and elastic' on the authority of a manuscript emendation made by Newton to his own copy of the second edition of *Principia*.) This became Newton's official position on the electric spirit. It was as far as he dared to go.

The culmination of his reasoning, however, came with two extraordinary drafts for the later queries of the *Opticks*, among the last scientific statements he ever made. The first notably added magnetism to the list of phenomena produced by the electric spirit:

> Do not all bodies, therefore, abound with a very subtil active vibrating spirit, by wch light is emitted, refracted, & reflected, electric & magnetic attractions & fugations [*i.e.* repulsions] are performed, the small particles of bodies cohaere when contiguous, agitate one another at small distances, & regulate almost all their motions among themselves as the great bodies of the Universe regulate theirs by the power of gravity?. ffor electric bodies could not act at a distance without a spirit reaching to that distance.

It also brought in a connection with the life force and the processes which drive living things, where it seemed to have an even stronger power than in

---

[19]Draft for General Scholium (MS C); *Unp.*, 361–362, 364.
[20]General Scholium; see Koyré and Cohen (1960).

inanimate ones:

> The vegetable life may also consist in the power of this spirit supposing
> that this power in substances wch have a vegetable life is stronger then in
> others & reaches to a greater distance from the particles. ffor as the electric
> vertue is invigorated by friction so it may be by some other causes. And
> by being stronger in the particles of living substances then in others it may
> preserve them from corruption & act upon the nourishment to make it of
> like form & vertue wth the living particles as a magnet turns iron into a
> magnet & fire turns its nourishment to fire & leaven turns past to leaven.
> ffor the living particles may propagate the vibrating motions of this spirit
> into into [*sic*] the contiguous particles of the nourishment & cause ye spirit
> in those particles to vibrate & act after ye same manner & by that action to
> modify the nourishment after the same manner with the living particles.[21]

The second follows on almost immediately, this time leaving out
magnetism:

> Do not all bodies, therefore, abound wth a very subtile active potent elastic
> spirit, by w[ch] light is emitted, refracted, & reflected, electric attractions
> & fugations are performed, & the small particles of bodies cohaere when
> contiguous, agitate one another at small distances, & regulate almost all
> their motions among them selves, ffor electric — uniting the thinking soul &
> unthinking body. This spirit may be also of great use in vegetation, wherein
> these things are to be considered, generation, nutrition & praeparation of
> nourishment.[22]

There has been some discussion about whether the electric 'spirit' is
aether or something contained within it, a kind of 'field' in modern terms.
In my view, it is almost certainly the latter, as this would fit in with
Newton's earlier ideas in which active 'spirits' were embedded within a
rarefied universal aether, and this view seems to be supported by a draft
query in which Newton writes: 'There are therefore Agents in Nature able
to to make the particles of bodies attract one another very strongly & to
stick together strongly those by attractions. One of those Agents may be
the Æther above mentioned whereby light is refracted. Another may be the

---

[21] Draft Q 18 B for *Opticks*, 1717, CUL, Add 3970, f. 241[r]; Home (1982, 202).

[22] Draft Q 25 for *Opticks*, 1717; McGuire (1968, 176); Westfall (1971, 394) and (1980, 793);
Home (1982, 198–200). These may be compared on the life force with draft Q 25: 'it may
be presumed that as electric attraction is excited by friction so it may be invigorated also
by some other causes & particularly by some agitation caused in the electric spirit by the
vegetable life of the particles of living substances: & the ceasing of this vigour upon death
may be the reason why the death of Animals is accompanied with putrefaction' (CUL,
Add. 3970.3, f. 235[v], NP).

Agent or Spirit which causes electrical attraction.'[23] The expression 'electrick Vapour' in Query 8, and the impact action of its force, also hint at a material, although highly rarefied, nature for the 'Spirit', as opposed to the seemingly non-material nature of the 'Æther'. In principle, however, it doesn't really matter, for either would still be true! The choice, in our terms, is effectively between nonlocal and local descriptions.

Whatever exact mechanism he contemplated, Newton was sure that

> ...the agent or spirit which glass emits by friction, is agitated with various motions like a wind, carrying along with it little light bodies by means of which you may know which way it moves. And since for the most part it carries those bodies towards the glass and makes them stick to it, we may thence conclude that it is the spirit by which electric bodies perform their attraction.[24] For though this agent acts at great distances (only) except when it is excited by the friction of electric bodies: yet it may act perpetually at very small distances without friction and that not only in bodies accounted electric, but also in some others.[25]

Newton had once again transformed a mass of conflicting ideas from a multitude of sources, experimental, theoretical and even non-scientific, into a recognisably scientific notion that still has validity in modern terms. However, many alchemists had previously talked about subtle and pervasive 'spirits', we know that Newton's 'active, potent, vibrating electric spirit, by which light is emitted' and 'electric & magnetic attractions . . . performed' has all the characteristics of something very like our own electromagnetic field, and seemingly, in the case of frictional electricity, one created by a highly rarefied collection of almost infinitesimally small material-like particles with electrical properties, not massively different from our modern electrons.[26]

7.3. Magnetism and Other Effects

There were some indications that Newton's electrical synthesis also included magnetism, or could do so, though Newton seemed to be extending himself again when he suggested a connection between the two forces. As we have seen, two drafts of the same late query are identical in nearly all respects except that one has the vibrating electric spirit responsible for both electric and magnetic effects and the other for electric effects only. It is as though he

---

[23]CUL, Add. 3970.3, f. 622^r, NP.

[24]Draft Observations for *Opticks*, c 1716; Home (1982, 207–208).

[25]Draft Observations for *Opticks*, c 1716; McGuire (1968, 180).

[26]Some key developments of a later period, stemming directly from Hauksbee's experiments with evacuated glass globes, led eventually to the discovery of cathode rays and the electron as a fundamental component of matter.

were still tentative about the connection, though his instinct told him that it must exist.

Hauksbee had found that electrical forces might be repulsive as well as attractive just like the forces between the particles of bodies. Magnetism had the same kind of properties and seemed to result from the magnetic nature of particles; possibly this also resulted from the same kind of cause. Newton also seemed to think that magnetism was to be found within bodies and that the particles of bodies were intrinsically magnetic. As well as the one definite claim in the draft query, that the electric agent was also responsible for magnetic attraction and repulsion, these forces were also clearly associated in other statements, which imply that magnetism is to be found within bodies and that the particles of bodies are intrinsically magnetic: 'All dense bodies seem to consist of electric particles and also to have a certain number of magnetic particles; as attraction of gravity suffices to explain the greater movements of planets and comets and sea, so electric and magnetic forces explain the interactions and movements of the particles of each several body.'[27] Now he seemed to be associating it, however tentatively, with electricity and denying its uniqueness as a property of iron alone.

Another draft implies that magnetism must be more general than it normally appears, possibly because 'iron' particles were contained in many bodies:

> Whoever therefore will have undertaken to explain the phenomena of nature which depend on the fermentation and vegetation of bodies and on the other motions and actions of the smallest particles among themselves will need especially to turn his mind to the forces and actions of the electric spirit, which (if I am not mistaken) pervades all bodies and to investigate the laws that this spirit observes in its operations. Then magnetic attractions will also need to be considered since particles of iron enter the composition of many bodies. These electric and magnetic forces ought to be examined first in chemical operations and in the coagulation of salts, snow, crystal, fluxes, and other minerals in regular figures and sometimes in the figures of plants, and afterwards to be applied to the explanation of the other phenomena of nature.[28]

The interest in symmetry, in life forms as well as in minerals, is also evidenced at the end of Query 31 of the *Opticks*.

Of course, the only form of magnetism then known, apart from that of the Earth, was ferromagnetism; and the Sun and Earth could not be ferromagnetic because of their heat. According to H. W. Turnbull, Newton's

---

[27] CUL, Add MS 3970, f. 240ʳ; McGuire (1968, 180).
[28] Draft connected with General Scholium; Westfall (1971, 393).

argument against the Sun being a magnet of the ordinary kind is now used to show that the magnetism of the Earth cannot be ferromagnetism.[29] This is because the Earth's core temperature exceeds the Curie point (1043 K), above which the spin orientations of the unpaired electrons become random. It is believed that the Earth acts instead as a self-generating dynamo and that its magnetism is produced by electric currents set up in a liquid iron outer core.

Earlier comments had emphasised the differences between magnetism and other forces: '...magnetick bodies when made red hot lose their vertue. A red hot iron loadstone attracts not iron, nor any Loadstone a red hot iron, nor will a loadstone propagate its vertue through a rod of iron made red hot in ye middle.'[30]

> Magnetic virtue is destroyed by a flame, and by heat: a rod of iron, either by standing long in a perpendicular position, or by cooling in an erect position, acquires magnetic virtue from the Earth. But it gets magnetic virtue too with a strong blow of a hammer at either extremity. If it is struck hard at one or other end the poles of the iron rod are interchanged: if it is struck in the middle (say with hammering at an anvil) it quite loses its magnetism. And so this virtue seems to be produced by mechanical means.[31]

Numerous drafts suggest an almost unified vision of the forces involved in electricity, magnetism and optics, and the cohesion of bodies, with special roles being played by vibrations and light:

> There is therefore an electric spirit by which bodies are in some cases attracted in others repelled & this spirit is so subtile as to pervade & pass through the solid body of glass very freely in both cases, & is capable of contraction & dilatation expanding it self to great distances from the electric body by friction, & therefore is elastic & susceptible of a vibrating motion like that of air whereby sounds are propagated, & this motion is exceeding quick so that the electric spirit can thereby emit light. And that which emits light in the experiments above mentioned, may emitt it in all shining bodies whenever sufficiently agitated either by heat or by putrefaction. And the Medium which emitts light may also be able to refract & reflect it as was noted above. This spirit may be also the Medium by whose vibrating agitations stirred up within dense bodies, the bodies receive heat & communicate it to contiguous bodies; the vibrations being propagated from one body into another where the bodies are contiguous,

---

[29] H. W. Turnbull, *Corr.* II, 362.

[30] Newton to Crompton for Flamsteed, 28 February 1681, *Corr.* II, 340–347.

[31] D. Gregory, *Annotations Physical, Mathematical and Theological from Newton*, 5, 6 and 7 May 1694, *Corr.* III, 338.

but reflected at the surface where they are not contiguous & by reflections kept within the hot body.[32]

Do not electric bodies by friction emit a subtile exhalation or spirit by which they perform their attractions? And is not this spirit of a very active nature & capable of emitting light by its agitations? And may not all bodies abound with such a spirit & shine by the agitations of this spirit within them when sufficiently heated? For if a long cylindrical piece of Ambar be rubbed nimbly it will shine in the dark & if when it is well rubbed the finger of a man be held neare it so as almost to touch it, the electric spirit will rush out of the Ambar with a soft crackling noise like that of green leaves of trees thrown into a fire, & in rushing out it will also push against the finger so as to be felt like the ends of hairs of a fine brush touching the finger. And the like happens in glass... And if the glass was held neare pieces of leaf brass scattered upon a table the electric spirit which issued out of the glass would stir them at the distance of 6, 8 or 10 inches or a foot, & put them into various brisk motions, making them sometimes leap towards the glass & stick to it, sometimes leap from it with great force, sometimes move towards it & from it several times with reciprocal motion, sometimes move in lines parall[el] to the tube, sometimes remain suspended in the air, & sometimes move in various curve lines.[33]

... the electrick spirit is so subtile as readily to pass through glass tho not so readily as through the Air. And whilst it pervades dense bodies so easily, why may it not be latent in them all in some measure or other, tho those only emitt it by friction in which it abounds most copiously? And since it easily emits light by agitation, why may it not emit light in all dense bodies heated red hot & thereby cause them to shine?[34]

The electric force may be invoked to explain cohesion:

May not the forces by which the small particles of bodies cohere & act upon one another at small distances for producing the above mentioned phænomena of nature, be electric? For altho electric bodies do not act at a sensible distance unless their virtue be excited by friction, yet that vertue may not be generated by friction but only expanded. For the particles of all bodies may abound with an electric spirit which reaches not to any sensible distance from the particles unless agitated by friction or by some other cause & rarefied by the agitation. And the friction may not rarefy the spirit not of all the particles in the electric body but of those only which are on the outside of it: so that the action of the particles of the body upon one another for cohering & producing the above mentioned phænomena may be vastly greater then that of the whole electric body to attract at a sensible distance by friction. And if there be such an universal electric spirit in body,

---

[32]Draft Q 25 for *Opticks*, 1717; CUL MS Add. 3970.3, 241^{v}; Home (1982, 198–200).

[33]Draft Q 18 B for *Opticks*, 1717; CUL MS Add. 3970.3, f. 293^{r}; Home (1982, 202).

[34]Draft Q 18 B for *Opticks*, 1717; CUL MS Add. 3970.3, f. 293^{r}; NP; Home (1982, 202).

certainly it must very much influence the motions & actions of the particles of the bodies amongst one another...[35]

Cohesion by electrical attraction of the parts was manifested by the spherical shape of liquid drops in the same way as the gravitational attraction of the parts created the spherical shape of the Earth.[36]

The electric agent is seemingly responsible for motions within bodies and in this way produces motions on a small scale similar to the large-scale motions of the universe produced by gravity. These motions in microscopic matter certainly seem to have been centripetal in character, and the calculations involved in refraction supposed forces of this kind within atoms. There are tantalising comparisons of the great and little motions: possibly, Newton imagines that the same sort of centripetal motions occurred on a small scale within atoms as on the large scale within the Solar System. We can only speculate on whether he had any vision of a planetary model for the component particles of matter. Several passages are very suggestive in this respect, including one in Hauksbee's *Physico-Mechanical Experiments*, which he may have inspired: '...so that in these smaller orbs of matter, we have some little resemblances of the grand phenomena of the universe.'[37] This was in response to Hauskbee finding that threads he attached to a disc placed inside an electrified globe, pointed inward to the centre of the disc in the manner of action of a centripetal force, though they were repelled when he brought his finger close to the globe.

Newton thought that, if we could understand the internal motions of the small particles of matter, we could also understand a large part of the phenomena of nature. This was a very difficult task, and at the time not much more than a programme. Somewhere inherent in the work so far accomplished, however, is the idea that all the results of forces or, as we should say, forms of energy, are equivalent and interchangeable, with heat as the agent of change; this is because the strengths and natures of the forces, as manifested, are a result of the configurations of the particles of matter within substances, and heat is the means by which changes in these configurations are made. A key concept is that the electric force, when liberated by friction, travels long distances, and so must be vibratory; light is vibratory because it is periodic, and so it must be electric vibrations which emit light. Heat is also vibratory and is present in all substances, so electric vibrations must

---

[35]Draft Q 24 for *Opticks*, 1717; CUL MS Add. 3970.3, f. 235ʳ; NP; McGuire (1968, 175–176); Westfall (1971, 315); Home (1982, 197–198).

[36]Draft Conclusion to the *Principia*, translated Cohen and Whitman (1999, 290).

[37]Hauksbee (1706, 1707); (1719, 54–55, 67, 74–75, 143, 154–155). Heilbron (1982, 169–170).

cause the vibrations of molecules of which heat consists. A curious passage in a draft Conclusion to the *Principia* seems to imply that heat alone can produce electrical action in amber and diamond: 'That spirit is also emitted from some bodies (as from electrum [i.e. amber] and adamant [i.e. the hardest substance, diamond]) by heat alone without friction, and attracts small light bodies.'[38] Diamond can liberate electrons by thermionic emission, but only under rather more extreme conditions than Newton would have encountered.

7.4. Chemistry and the Life Force

Beyond the purely physical, chemical effects are produced by attractive forces of the same kind and so they too must be electric. As Newton informed the Royal Society from the Presidential chair in 1710: 'most of the Phaenomena appearing in Fermentations, Dissolutions, Precipitations, and other Actions of the small Particles of Bodies one upon another were caused by Electrical Attraction'[39], an insight which was not to be properly followed up for the next 100 years. This verbal announcement was the most startling of Newton's statements on chemistry, and was probably related to the fact that the chemical operations in nature were due to the largest bodies in the hierarchy of particles and the electrical force was the first to be extracted from matter, and merely by friction of its outer parts; but, whatever its origin, it significantly predates the discovery of the chemical generation of electricity in the Voltaic cell. That chemistry is due to the electrical force relates also to the electrochemical series of affinities in metals, which Newton published in the optical Queries of 1717, though without specific reference to the fact that he already knew that the affinities could be treated as electrical forces.

Electricity, as we have seen, would seem to be responsible for the life force as well. In *De Vita & Morte Vegetabili*, a draft related to the queries, and in a draft preface for a new edition of the *Principia*, Newton sees the electric force or 'spirit' as responsible for the vitality of living organisms.

> The particles of bodies coalesce and cohere in diverse ways by means of the electric force.... And [as] by the friction of an electric body its attraction is extended, so also by the action of living the electric force of living parts is extended, and by a strong attraction it happens that those parts both conserve their proper form and situation and impart their nutriment by degrees in the same manner as a magnet converts iron into a magnet and fire converts bodies into fire and leaven converts paste into leaven:

[38] Draft Conclusion to the *Principia*, translated Cohen and Whitman (1999, 288).
[39] *Royal Society Journal Book*, 19 April 1710. Heilbron (1982, 171).

but with the ceasing of vegetable life that vital attraction ceases and its absence immediately begins the action of dying which we call corruption and putrefaction.[40]

The suggestion here and in the passage, already quoted, beginning 'The vegetable life may also consist in the power of this spirit,' that the replication of biological tissues through the electric force is similar, as a process, to the transfer of what we would call magnetic ordering is particularly striking. Both reflect fundamental dualities in nature, the genetic code, which is the ultimate source of biological replication, being created from the double strands of DNA, and magnetism being intrinsically dipolar; in each case, it is the attraction of the dual partner that creates the extension of the order and structure. There are also generic similarities between the 'replication' process described in this passage and one producing such things as metals in the Earth, which Newton had discussed earlier in *Of Natures Obvious Laws & Processes in Vegetation*, and which was already being described by a metaphor analogous to the life process,[41] while 'magnetic' metaphors had appeared in other alchemical writing from the period. It may be that the biological concept is a remote descendant of the earlier 'alchemical' way of thinking, derived by Newton's usual method of progressive abstraction.

In a draft Conclusion to the *Principia*, he specifically says that the electric force causes living bodies to absorb nourishment by attracting 'parts of nutriments similar to themselves,' while, in drafts for the *Recensio libri*, he connects the electro-optical agent with that of 'sensation', and imagines the electric spirit as acting upon that substance in the 'sensorium' 'which sees and thinks' and is responsible for animal motion.[42] In more general terms, he says:

> And because the force of gravity is so widespread that all the movements of the heavenly bodies are governed by it, and our sea too is caused to ebb and flow, it remains to investigate the remaining attractive forces, that is to say the electric force and the force of magnetism, and to discover their laws and their effects upon the motions of the least of bodies in solution,

---

[40] *De Vita & Morte Vegetabili*; Hawes (1971). Iliffe (2004) says that 'Harking back to his work on the vegetation of metals,' he suggested that this force [the electrical] was implicated in the union of soul and body, as well as in generation, nutrition and the 'preparation of nourishment.'

[41] *Of Natures Obvious Laws*, NPA. The term 'vegetation' in alchemical texts, of course, meant something far wider than the purely biological usage, but the metaphor is still suggestive, as is the one of 'trees' used in a metallic context.

[42] Draft Conclusion for the *Principia*, translated Cohen and Whitman (1999, 292); drafts for the *Recensio Libri* (1722), quoted Cohen (1999, 280, 281).

fermentation, vegetation and similar processes. For the effects of these forces are widely diffused..., and the electric force (as I perceive) is extended I might almost say through all bodies...[43]

Many passages about nerves and animal motion (often based on his own dissections) are found in Newton's official writings as well, though there 'vibrations' are discussed without being specifically described as electric in origin. In Query 12, vision is the held to be the result of the propagation of waves excited by the rays of light in the optic nerves; because dense bodies conserve their heat a long time, and the densest bodies conserve their heat the longest, the vibrations of their parts are of a lasting nature, and therefore may be propagated along solid fibres of uniform dense matter to a great distance, for conveying into the brain the impressions made upon all the organs of sense. According to Query 14, the harmony and discord of colours arises from the proportions of the vibrations propagated through the fibres of the optic nerves into the brain, as the harmony and discord of sounds arise from the proportions of the vibrations of the air; some colours, if viewed together, are agreeable to one another, as those of gold and indigo, others disagree. Query 15 discusses binocular vision, along with other aspects of seeing, suggesting in particular a concept of semi-decussation or rearrangement of the fibres of the optic nerves (in the optic chiasma), such that those on the left side of each nerve unite in a single place and go to the brain from the tract on the left side of the head, with those on the right side of each nerve going to the brain from the tract on the right side of the head. So, each of the optical tracts contains half the fibres from each optic nerve, covering both left and right halves of the visual field. Query 16 describes the persistence of vision, a subject on which Newton had dangerously experimented as a young man, with himself as subject; the persistence is lasting because of vibrations.

In his manuscript treatise *De vi Electrica*, Newton describes the physiological action of the eye, and the other sense organs, together with the operation of muscle, as being the result of electrical impulses transmitted along the nerves:

> When the rays fall upon the bottom of the eye they excite vibrations in this medium which, carried through the solid capillaments of the optic nerve into the brain, create vision. And just as the vibrations excited in the air by resonant bodies produce all the tones of sound according to their magnitude and number, so the vibrations of the spirit hidden within the capillaments

---

[43]Draft Preface to new edition of *Principia*, Autumn 1712?, *Corr.* V, 113–114.

of the optic nerve create all the colours of light according to their magnitude and number. . . . And because nature is simple and conformable to herself, and this very spirit is the vehicle for transporting the impressions of light into the brain, the same must be the vehicle for transporting the impressions of sounds, odours, tastes and touch through the capillaments of other nerves into the brain, as also the impressions made by the will in the brain (through the capillaments of other [motor] nerves) into the muscles, in order to move the limbs of animals.[44]

Further detail is given in a draft Query:

The like vibrations may be excited in the bottom of the eye by light & propagated thence through the solid capillamenta of the optick nerves into the sensorium for causing vision & the like of other senses. The like vibrations may be also propagated from the brain through the solid fibres of the spinal marrow & its branches into ye muscles for agitating & expanding the liquors therein & thereby contracting the muscles to cause the motions of animals. For as [sic] liquors are expanded by heat & by consequence by the vibrating agitations of this spirit. If the agitations be of short continuance they expand the liquors without heating them for want of time to do it. If lasting (as in running a race, as in supporting a burden without external motion of the body) they heat the body by degrees & at length excite sweat. The spirit therefore may be the medium of sense of animal motion & by consequence of uniting the thinking soul & unthinking body.[45]

And in a manuscript draft preserved by Brewster, which also invokes an aethereal mechanism:

. . . Light seldom strikes upon the parts of gross bodies (as may be seen in its passing through them); its reflection and refraction is made by the diversity of ethers, and therefore its effect upon the retina can only be to make this vibrate, which motion must then be either carried in the optic nerves to the sensorium or produce other motions that are carried thither. . . . However, what need of such spirits? Much motion is ever lost by communication, especially betwixt bodies of different constitutions. And therefore it can no way be conveyed to the sensorium so entirely as by the ether itself. Nay, granting me but that there are pipes filled with a pure transparent liquor passing from the eye to the sensorium and the vibrating motion of the ether will of necessity run along thither. For nothing interrupts that motion but reflecting surfaces; and therefore also that motion cannot stray through the reflecting surfaces of the pipe, but must run along (like a sound in a trunk)

---

[44] *De vi Electrica, Corr.* V, 365–368.
[45] Draft Q, NP, Add 3970, ff. 241ᵛ–241ʳ.

entire to the sensorium. And that vision thus made is very comfortable to the sense of hearing, which is made by like vibrations.[46]

In the modern theory, worked out in the early to mid-20th century, the nervous system requires a particular kind of cell variously called a neuron, neurone, or nerve cell. These transmit electrical impulses along protoplasmic extensions called axons, which are packaged together in the fibrous bundles we call nerves. The impulses from the neurons are then passed to other cells at junctions across their membranes known as synapses. The membranes of all animal cells produce electrical impulses by pumping in two potassium ions at the same time as three sodium ions are pumped out. If this happens in a neuron, a pulse or wave of electrical activity will be transmitted along the axon to the synapse. Different kinds of neuron produce different kinds of activity in the human body. In the case of vision, the light entering the eye is focused by the lens onto the retina with its layers of nerve cells and ultimately along the axons of the optic nerve to the visual cortex, or optical processing part of the brain. In the case of muscular motion, the muscles move by receiving electrochemical signals from motor neurons to the brain.

Newton's descriptions of both vision and muscular motion, while seemingly not particularly profound, are correct in every respect. This is because, once he has decided on the electrical origin of the nerve impulse, there is no further assumption needed to create a completely adequate working description of the process. His method of theorising is almost designed to be fool-proof. As in other cases, minimalism leads to a completely successful generic description. He extracts the components of the theory which he clearly thinks are obviously correct and adds nothing extra which would need to be more hypothetical. The key breakthrough, perhaps influenced partly by the significance of the hand's action on Hauksbee's 'plasma ball' and in his own earlier experiments, lies in his realisation that the electrical force can be applied to animal physiology along with many other biological and chemical phenomena. The detailed mechanism could be left to be worked out when experimental techniques became advanced enough to make the necessary investigations.

Query 23 again describes vision as resulting from vibrations of the aetherial medium, transmitted via the 'optick Nerves'; and suggests the hearing and the other senses result from analogous processes. In Query 24, as in *De vi Electrica*, animal motion is posited as resulting from vibrations in the medium sent from the brain 'by the power of the Will,' and propagated

---

[46] Manuscript draft, printed in Brewster (1855, I, 435–436).

through the nerves to the muscles, allowing them to contract and dilate. The medium referred to in these queries seems to be clearly identifiable with the 'spirit' of the General Scholium.

An extraordinary cancelled passage from an unpublished Preface to the second edition of the *Principia* hints at something even more startling. After he has spoken about the promotion of philosophy through investigation of 'the laws and effects of electric forces,' beginning with the phenomena and proceeding to their causes, Newton says: 'And the actions of the mind of which we are conscious should be numbered among the phenomena.'[47] In a preliminary draft he had written: 'What is taught in metaphysics, if it is derived from divine revelation, is religion; if it is derived from phenomena through the five external senses, it pertains to physics; if it is derived from knowledge of the internal actions of our mind through the sense of reflection, it is only philosophy about the human mind and its ideas as internal phenomena likewise pertain to physics.'[48] How extraordinary it is — and even shocking — to read a passage like this from someone like Newton, effectively rejecting the Cartesian dualism and the ghost in the machine, even if we take into account his professed antipathy for the work of Descartes![49]

And yet, this was essentially the logical conclusion to Newton's rejection of the Cartesian philosophy almost 50 years earlier. Newton regarded his views of the relation between God and nature as described in the *Principia* and *Opticks*, as counteracting the 'atheism' he perceived in Descartes, which arose from the latter's assumption that extension was matter. Matter, for Newton, was not everywhere, and consequently could be conceived independently of God's 'boundless uniform Sensorium', whereas extension, which was everywhere, could not. For Newton, minds, like bodies, but unlike the omnipresent and universal deity, had to occupy a place, and objects occupying a place would have all the characteristics of bodies.[50] As he put it in his only metaphysical work, *De Gravitatione et Aequipondio Fluidorum*, written at an earlier date and specifically directed against Descartes: 'God

---

[47] Cancelled Preface to the *Principia*, translated Cohen and Whitman (1999, 49–54).

[48] Draft for cancelled Preface to the *Principia*, translated Cohen and Whitman (1999, 49–54). A similar argument appears in the proposed Rule V intended for the *Principia* (see *Newton and the Great World System*).

[49] Newton's views on the mind-body relation necessarily required the 'heresy' of mortalism. See Dempsey (2006); also Force (1994, 1999); Smolinski (1999); and Snobelen (2004). Gorham (2011, 32) says that Newton's 'solution to the mind-body problem' in *De Gravitatione* 'anticipates George Berkeley, while relying on a theory of causation at the heart of Cartesian metaphysics.' Kochiras (2013) presents a counter-argument.

[50] See, for example, Stein (2002, 271–272).

is everywhere, created minds are somewhere, and a body is in the space it fills.'[51]

Newton's metaphysical views, of course, led him into unorthodox theology, such as the denial of the Trinity, and intimations of this may well have led to some of the attacks on his natural philosophy in his later years, as well as accusations of 'atheism' against Newton himself — an astonishing reversal. They also meant that created minds could interact in the same way as bodies through forces acting at a distance, or 'active powers', which emanated directly from the greater Sensorium of God, and which ensured that the preservation of activity in the world (effectively, our conservation of energy) did not require the perpetual motion of matter, as Descartes and Leibniz had assumed, or the conservation of the scalar value of 'motion', or, in our terms, momentum (which Newton saw as 'always on the decay'). Newton's desire to counteract the compromising tendencies in Descartes' works which implied limitations on the absoluteness of the deity had led to a new philosophy of mind as well as of matter.

In some sense all of nature was animate. As he wrote in a draft Query: 'We cannot say that all Nature is not alive.'[52] And in another:

> And since all matter duly formed is attended with signes of life & all things are framed with perfect art & wisdom & Nature does nothing in vain; if there be an universal life & all space be the sensorium of a thinking being who by immediate presence perceives all things in it as that which thinks in us perceives | see their pictures in the brain: the laws of motion arising from life or will may be of universal extent.[53]

In the classical scholia of the 1690s, he had argued that the early Ionian philosopher 'Thales regarded all bodies as animate, deducing that from magnetic and electrical attractions. And by the same argument he ought to have referred the attraction of gravity to the soul of matter....'[54] Thales had seen that even material bodies had something which made them seem to act consciously. Newton was now sensing that it was actually such properties which were ultimately responsible for the consciousness.

Alan Gabbey refers to the 'unproblematic' nature of mind–body interactions in Newtonian theory, and quotes a passage from *De Gravitatione* where Newton had imagined God willing a specific region of space to be impenetrable and so creating something with all the characteristics of a

---

[51] Stein (2002), including translation 269.
[52] Draft Q 23, CUL, MS Add. 3970.3, f. 620^v, NP.
[53] Draft Q 23, CUL, MS Add. 3970.3, f. 619^r, NP.
[54] Draft Scholium to Proposition 9; McGuire and Rattansi (1966).

material particle. Equally God, by a similar act of will, could stimulate human perception to do the same. Because the Newtonian philosophy had no such constraining principle as the conservation of *vis viva* or the maintenance of perpetual motion in the world, Newton's followers in the 18th century, working in physiology, psychology and the theory of mind, were able to develop their ideas freely without the constraints imposed on physical mechanisms, so giving Newton a 'recognized', if 'unexpected', 'place in the twin histories of psychology and philosophy of mind.'[55]

In *De Gravitatione*, Newton had argued against bodies 'having in themselves a complete and independent reality,' in preference to ascribing reality to the 'kinds of Attributes which are real and intelligible in themselves and do not require a subject in which they inhere,' for example, space, which could be imagined without substance in a vacuum. 'In the same way,' he continued (in Howard Stein's translation), 'if we should have an Idea of that Attribute or power by which God, through the sole action of his will, can create beings: we should perhaps conceive that Attribute as it were subsisting of itself, without any substantial subject, and involving his other attributes.' As long as we are unable to 'form an Idea of this Attribute,' or even of the 'power by which we move our own bodies,' 'it would be rash to say what is the substantial foundation of minds.' So, for Newton, we could even conceive of God 'entirely in terms of his attributes,' if we knew what they were. According to Stein, Newton has here set aside the Cartesian 'mind–body dualism or monism, in favour of the program: *to seek to understand mental attributes and their relation to corporeal ones.*'[56] Perhaps Newton now felt he had glimpsed the possibility by which this understanding could be reached. It would be entirely consonant with his characteristic tendency to increasing abstraction in his physical explanations.

Stein also points out that bodies, for Newton, are endowed 'with the power to *interact with minds*,' and 'that they are able, when they form part of what he calls our 'sensorium' [the crucial region of our brain], to induce

---

[55]Gabbey (2002, 329–357). I don't believe in Gabbey's implication that there was a 'mismatch between the third law [of motion] and [Newton's] inviolable belief in the power of the human mind to intervene in the mechanism of the world,' as the third law is a conservation principle of *vector* momentum, and not constraining in the same way as a scalar conservation law would have been. It also allowed Newton to introduce 'active principles', leading to a different kind of scalar conservation law, and ultimately to the mathematical realisation of the nature 'on the decay' concept in the nonconserved quantity of entropy. Interestingly, it is in living systems, and probably ones with a high degree of 'mind' or consciousness, that entropy tends to be at a maximum.

[56]Stein (2002), including translation, 281–282.

specific forms of *awareness* as a consequence of specific motions on their part; and, correspondingly, that our acts of will cause suitable motions in those that initiate activity in what we now call our motor neurons,'[57] exactly as Newton was proposing for the electrical force. Newton himself writes, in *De Gravitatione*, of 'a faculty in bodies by which they are capable of union through the forces of nature. From the fact that the parts of the brain, especially the finer ones to which the mind is united, are in a continual flux, new ones succeeding to those which fly away, it is manifest that that faculty is in all bodies.'[58] Newton's fundamental metaphysical principles, were never assumed to be *a priori*, but as the best available that could be derived from experience[59]; and these included the nature of the relationship between mind and body. Now, he had a mechanism suggested precisely by such experience. It was entirely characteristic of Newton, in his 'recursive' mode, that he should push his physical intuition to the very limit that anyone in a God-fearing age would conceivably dare to go.

7.5. The Laws of Electric and Magnetic Attraction

The various forces which Newton supposed to exist within the observable parts of matter were not, for him, independent, and it is not surprising that he eventually combined them all into a single electrical force. The subtle vibrating electric spirit explained all electromagnetic phenomena, light, heat, cohesion, capillary action, surface tension, and chemical and biological processes, all, that is, except gravity and the operations of the 'most secret works of Nature.' In his final descriptions, left out of the *Opticks* at the last minute in favour of a less controversial treatment of an electrical aether, and only published in the most cryptic form in the second edition of the *Principia*, the agent is subtle (not ordinary matter but capable of permeating matter), potent (powerful or, as we should say, energetic), vibrating, elastic, and present in all bodies; it causes the emission, reflection, refraction and diffraction of light, the transfer of heat, electric and magnetic attractions and repulsions, the cohesion of bodies, and all chemical and biological processes. The vibrations of this agent produce all the wave effects of light, and the vibratory motions associated with heat. And the agent is in itself electric, though more powerful than the ordinary electric

---

[57]Stein (2002, 280).

[58]*De Gravitatione*, translation by Stein (2002, 281).

[59]Stein (2002, 261) stresses that Newton's view's on God's relation to nature come at the *end* of both the *Principia* and the *Opticks*, as consequences of the results of his scientific investigations, not as assumptions made prior to carrying them out.

force induced by friction, although this itself is much more powerful than gravity. This is because it is hidden within bodies and difficult to remove from them; only the effect of friction, breaking the particles of matter into something yet 'more subtle' than air (a fourth state of matter) 'by some violent action,' allows it to be forced out of them, and only then in a relatively small degree. But the enormous strength of the electric interactions within bodies can be seen from the very powerful nature of the forces of cohesion and capillary action, already calculated.

Newton was clear that the electric force and probably the magnetic force must be prevalent at the micro level in matter, and must be of a similar nature to gravity in some respect. But what was the mathematical form of the laws relating to these forces, that is, how did they vary with distance? Some indication of the type of force law associated with electricity could be found if it was associated with capillarity, as investigated by Hauksbee in the experiment on oil of oranges. Curiously, Newton's very first discussion of capillary action or, as he called it, 'filtration', in the early *Quaestiones Quaedam Philosophicae*, is in a section with the title 'Attraction Electrical & Filtration,' although electricity is not discussed there.[60] In Hauksbee's experiment, there was some kind of inverse square relationship, though not exactly the inverse square relationship between static charges known to us. Newton writes: 'It is clear also that this force is by far strongest at the very surface of the glass. In the latest experiment, the force of attraction came out very nearly inversely in the ratio of the square of the distance [i.e. very nearly as the inverse square of the distance] of the drop of oil of oranges from the concourse of the glasses.'[61]

A version of this is in the published Query 31:

Now by some Experiments of this kind, (made by Mr. *Hauksbee*) it has been found that the Attraction is almost reciprocally in a duplicate Proportion of the distance of the middle of the Drop from the Concourse of the Glasses, *viz.* reciprocally in a simple Proportion, by reason of the spreading of the Drop, and its touching each Glass in a larger Surface; and again reciprocally in a simple Proportion, by reason of the Attractions growing stronger within the same quantity of attracting Surface. The Attraction therefore within the same quantity of attracting Surface, is reciprocally as the distance between the Glasses. And therefore where the distance is exceeding small, the Attraction must be exceeding great.

---

[60]QQP; 'Attraction Electrical & Filtration,' f. 103, NP.
[61]Draft Conclusion to the *Principia*, translated Cohen and Whitman (1999, 287–292).

Newton realises that the force of surface tension decreases with the linear distance over which it acts. In our terms, this is because the energy is the same for each molecule and so the total energy is proportional to the area of surface. That is, the inverse proportionality between force and distance for capillary action at small separations will be equivalent to a constant value for force × distance, or energy of molecular interaction. (A constant value for the intermolecular potential energy in *gases* is similarly implied in Newton's derivation of Boyle's law in the *Principia* (Book II, Proposition 23).) For a molecule of dimension R and cohesive energy U, the energy per unit area at the surface will be $U/2R^2$, the other half being cancelled out within the liquid. As Newton also realises, for an oil drop treated as a cylinder, the area will increase as the thickness of the drop decreases, and, hence, in the Hauksbee experiment, the force will increase in another inverse proportion to thickness or distance, producing a total force inversely proportional to the distance squared. It works out, in effect, as: force $\propto$ energy/distance $\propto$ area/distance $\propto$ constant volume/distance^2.

In modern experiments, the respective surface tensions of oils, water and mercury are measured at 0.02, 0.07 and 0.5 Nm^{-1}. The result Newton calculates from the Hauksbee experiment,[62] in which a column of height 10^{-7} inches supports a weight of 5.55×10^{-11} N, gives an excellent value for the surface tension of oil of oranges, or the oil/glass boundary, of 0.022 Nm^{-1} (or, 0.02185 Nm^{-1}, in more exact terms). This is the only direct measure of a force of electrical origin that was available to Newton. If we take 0.022 Nm^{-1} as the force per unit length (F/R) from the experiment over the distance $R = 10^{-7}$ inches $= 2.54 \times 10^{-9}$ m, then, using the modern value of $4\pi\varepsilon_0 = 4\pi \times 8.85 \times 10^{-12}$ Fm^{-1}, we have a value of the equivalent of charge (e) squared $= 4\pi\varepsilon_0 \, FR^2 = 4\pi \times 8.85 \times 10^{-12} \times 0.02185 \times 2.54^3 \times 10^{-27}$C^2, leading to a value of 2×10^{-19} C for the equivalent charge, a truly remarkable result. Because the experiment produces an inverse square dependence of the force in the configuration used, the result is not sensitive to the height of the column, and could, in principle, be reduced to the length over which a single charge acts.

From the density (840 kgm^{-3}) and chemical formula ($C_{10}H_{16}$) of d-limonene (molar mass 136.23), we may estimate the separation of the individual oil molecules as 0.65×10^{-9} m, and perhaps approaching 1.0×10^{-9} m for their longest dimension, which is only slightly smaller than the maximum of 10^{-7} inches which Newton had derived from the separation of contiguous

---

[62] *De vi Electrica, Corr.* V, 368.

glass slides, showing how close he was to estimating the true value of the molecular scale in matter. It would mean that Newton's estimate of the smallest unit to display the oil's physical properties would contain about 50 molecules, 3 or 4 in each direction.

Using the separation value for the molecular diameter would make the potential energy for a single molecule of d-limonene close to 0.06 eV but we can assume that Newton would have equated summations of the optical and capillary forces over similar distances. At 10^{-7} inches, using Newton's values for force leads to a potential energy of $5.55 \times 10^{-11} \times 2.54 \times 10^{-9} = 1.4 \times 10^{-19}$ J or 0.88 eV, which is interestingly close to the electromagnetic energies between 1.8 to 3 eV in the visible region of the spectrum. According to further calculations on f.622 of Cambridge University Library MS Add 3970, the equivalent potential energy for water at 10^{-7} inches would be greater than 1.3 eV, and so even closer to the optical energy range.[63]

One question that seemingly could be decided by experiment was whether the analogy of the electric and gravitational forces allowed the electric force to be transmitted, like gravity, through vacuum. Hauksbee had left this question undecided, following his experiment with the threads, being inclined, with the earlier electrical investigator Niccolo Cabeo, to reject vacuum transmission and see, rather, the action of chains of glass effluvia directing the threads. Newton, however, in the Latin *Opticks* of 1706, had implied that forces acting at a distance were responsible for all physical phenomena. Hauksbee's successor as Royal Society Demonstrator, John T. Desaguliers, was called upon to make the decisive test. An Oxford graduate, Desaguliers was more of an independent scientist than his predecessor and the work he did at the specific request of Newton is easily distinguishable from his own, more technologically-based research programme. On this occasion, in May 1717, in connection with the forthcoming edition of Newton's *Opticks*, he performed an experiment to show that the electric force did indeed pass through vacuum. At this point, Newton considered giving the electric force a prominent place in the *Opticks*, and was writing the draft Queries about the universal nature of the electric spirit. At the last minute, perhaps under pressure from the Cartesians against his supposed espousal of 'occult' qualities, he decided against this. Desaguliers makes a reference to the electric universe in *Physico-Mechanical Lectures*, an 80-page syllabus for a course of lectures that he published in the same year, but Newton ended the *Opticks* abruptly with a Book III, Part 1 and no Part 2,

---

[63] f. 622, Westfall (1971, 423).

before the 31 Queries, and also no mention of electricity as a fundamental force or of an electric spirit responsible for its actions. Instead, there was the introduction of a new speculative 'non-material and non-mechanical' aether, responsible for the transmission of both electric and gravitational forces, and presumably able to allow particles to interact at a distance without *mechanical* intervention.[64] In this way, also, the new Queries 17–24 could be reconciled with Queries 25–31, originally from 1706, which resolutely argued in favour of short-range forces between the particles of matter being responsible for all physical effects other than gravity, and which were extended and placed at the end of the book to emphasise the continuing importance of the concept.

The nature of the force law for magnetic bodies proved to be equally difficult to investigate, but also led to results of fundamental interest, which could have produced an even more significant breakthrough if it had come earlier in Newton's career. Again, like the electric force, it was not proportional to the quantity of matter in which it was situated. Newton's definition of the 'absolute' value of centripetal force in the *Principia* (essentially the value of the force as it derived from the source, rather than from its effect on the object acted upon) had distinguished gravity and magnetism by saying that different loadstones exerted forces that were not dependent on their 'bulk', but according to some other property.[65] It was also possible to change the strength of a magnetised object by various physical actions. It was, however, a force in which, like gravity and electricity, the total effect on any body was derived by adding up the effects on all the individual particles or source elements — 'attraction towards the whole arises from attraction towards the several parts' — and the Newtonian search for a force law was, therefore, for the one between these elements and not the one between the macroscopic bodies on which the measurements had to be made.[66] The distinction would prove to be a difficulty in pinning down the force law for a magnetic element, as magnetism was more complicated than gravity in having repulsion as well as attraction, and in having south poles as well as north poles.

Concerning the distance relationship, Newton's own early experiments had suggested that the magnetic force might follow an inverse-cube law

---

[64] *Royal Society Journal Book*, 9, 16, 23 May 1717. Desaguliers (1717), Appendix. Heilbron (1982, 170–171). For 'nonmaterial and non-mechanical', see Heilbron, 46, n. 33, which cites Heiman and McGuire (1971, 244).

[65] *Principia*, I, Definition 6.

[66] *Principia*, III, Proposition 7, Corollary 1. Heilbron (1982, 82). The Scholium after I, Proposition 69, specifically mentions magnetism in this connection.

rather than inverse-square. In the first edition of the *Principia*, of 1687, he wrote: '...in receding from the magnet it surely decreases in a ratio of the distance greater than the duplicate; because the force is far stronger on contact than is the case with attracting bodies which are very little separated from their neighbours.'[67] He resolved to investigate the problem when the opportunity came.

Inverse-cube laws were a special interest of Newton's from an early period. An inverse-cube law implied a relationship in which the force would be vastly greater at a small distance from the attracting object, suggesting how such forces could seem to be virtually confined within matter. Already, in the *Principia*, he had proved theorems to this effect, with the suggestion that he had made explicit elsewhere, that such a force (or one with an even greater inverse-power dependence) would be needed to explain cohesion. In addition, he had shown that an inverse-cube attraction between a large body and a smaller 'satellite' one would lead to rotation as the smaller body spiralled into the larger, while magnetism in any situation where there is an asymmetry between the fields around the attracting or repelling bodies naturally leads to rotation.

Now, in 1712, Newton, working closely with Roger Cotes, probably his most brilliant disciple, on the forthcoming second edition of the *Principia*, found that just such a law applied to the tides. The tide raising powers of both Sun and Moon were shown to vary with the inverse cube of their distances from the Earth. Newton and Cotes derived a general principle that the height of the tidal bulge due to a satellite on a planet is proportional to the inverse third power of the separation between satellite and planet, because the tide is created by a difference in the attractive force due to the satellite on opposing sides of the planet.[68] Effectively, the tidal force is the result of a dipolar action by the Sun or Moon on the Earth's seas, with each 'pole' subject to an inverse square law, but the sum of the forces inverse cube. The method of images used in Proposition 82 may also be associated with the dipole concept, while Corollary 6 to Proposition 44 could be regarded as the action of an inverse-cube dipole on a body without an elliptical orbit sustained by a monopolar centripetal attraction. Magnets had long been known to be dipoles and here was the opportunity to explain his previous magnetic result, and to extend the idea to cohesion. An even greater opportunity occurred almost simultaneously.

---

[67] *Principia*, III, Proposition 6, Corollary 5; Bechler (1974, 197).

[68] *Principia*, III, Propositions 36 and 37.

As President of the Royal Society, especially during the years 1703–1717, Newton was very much the leader of a research team, which included Cotes and Samuel Clarke on the theoretical side, along with Hauksbee and Desaguliers on the experimental, while Brook Taylor contributed to both. All of these were outstanding mathematicians, philosophers or physicists in their own right. While Cotes and Newton collaborated extensively on the second edition of the *Principia*, Hauksbee, Taylor and Desaguliers performed experiments, sometimes (as we know) at his suggestion or with his continued analysis and interpretation. The details of the collaborations are only known in a few cases from stray comments, but we can imagine that it was more extensive than now appears. The years of these collaborations were also those of the two controversies with Leibniz and his supporters, over the invention of the calculus and the validity of Newton's dynamics and world system, in which Newton had further supporters, notably John Keill.

The timing of events during the spring and summer of 1712 is quite extraordinary. Newton proposed in March 1712 'that Dr. Halley and Mr. Hawksbee, should try the power of the great loadstone at several distances to find the true proportion of its decrease, which he believed would be nearer the cubes than the squares'[69]; experiments were reported in the *Royal Society Journal Book* throughout the spring. On 22 April 1712, he wrote to Cotes: 'I am satisfied that the Moon upon the Sea is in triplicate ratio of her distance reciprocally.'[70] The first results on the magnetic experiment were reported on 25 June. Hauksbee and Taylor found that the magnetic attraction between the Society's lodestone and a compass needle is inverse-square close in but inverse-cube at greater distances.[71] Taylor ultimately concluded in a paper not published until 1721 that 'at the distance of nine feet, the power alters faster, than as the cubes of the distances, whereas at the distance of one and two feet, the power alters nearly as their squares... From whence it seems to appear, that the power of magnetism does not alter according to any particular power of the distances, but decreases much faster in the greater distances than it does in the near ones.'[72]

In the *Principia*, published in 1713, Newton wrote:

> The power of gravity is of a different nature from the power of magnetism;
> for the magnetic attraction is not as the matter attracted. Some bodies are

---

[69] *Royal Society Journal Book*, X, 375, March 1712. Armitage (1966, 75); Palter and Hynd (1972, 547).

[70] *Corr.* V, 273.

[71] *Royal Society Journal Book*, 25 June 1712. Taylor (1715b); Palter and Hynd (1972, 547).

[72] Taylor (1721).

attracted more by the magnet; others less; most bodies not at all. The power of magnetism in one and the same body may be increased and diminished; and is sometimes far stronger, for the quantity of matter, than the power of gravity; and in receding from the magnet decreases not as the square but almost in the triplicate proportion of the distance, as nearly as I could judge from some rude observations.[73]

In the same work, he wrote of the tide-raising powers of both Sun and Moon: '...the sea is raised not only in the places directly under the Sun, but in those which are directly opposed to it; and the sum of these forces... is... as the cube of the distance from the Earth reciprocally.'[74] 'The force of the Moon to move the sea is to be deduced from its proportion to the force of the Sun, and . . . is in the reciprocal triplicate proportion of its distance. . .'[75]

Hauksbee and Taylor's experiment had provided a massive clue for a new synthesis, and it is difficult to believe that a younger Newton would not have seized the opportunity. Newton's work with Cotes had shown that the action by the Sun or Moon on the Earth's seas was effectively dipolar, with each 'pole' operating as the source of an inverse square law, but with the sum of the forces becoming inverse cube. This is exactly what happens in Taylor's experiment on magnetic force. Close in, the interaction is monopole on monopole, and so inverse square, further out it becomes monopole on dipole, and so inverse cube. We have no evidence that Newton thought of applying the principle there, or to the inverse powers of distance greater than 2 which he recognised must apply to the cohesive forces in material bodies. At this moment, the opportunity existed for Newton to explain all known forces in terms of inverse square laws which became modified by the conditions in which they were applied — exactly as happened with perturbations in planetary orbits. He would also have had the perfect explanation for the screening and shielding of micro-forces within matter, which he had often commented upon, if electric attraction and repulsion had been attributed to dipoles in the same way as the magnetic variety. There is no evidence whatsoever that he took it, or that anyone else did from the published accounts in the *Principia* and *Philosophical Transactions*.

When these discoveries were published, Newton was over 70 years old and nearing the end of his active scientific career. The inverse-cube law of

---

[73] III, Proposition 6, Corollary 5.
[74] III, Proposition 36.
[75] III, Proposition 37.

tides seems to have owed as much to Cotes as it did to him. He was also embroiled in the increasingly bitter dispute with Leibniz and his supporters, and had many drafts to work through for the next edition of the *Opticks*. Essentially, it was a late development and there was probably insufficient time for the analogy to develop in his mind. It would be one of the problems that he left unresolved, never quite fitting the pieces he possessed into a new synthesis.

Convinced, however, that the future lay with an intensive investigation of the electric force, Newton is said by Martin Folkes, a Fellow of the Royal Society from 1714, and Secretary from 1722 to 1723, to have urged the Fellows on several occasions to continue the study. His comments are recorded only in a book written in French in the 1740s by the Comte de Tressan, and published in 1786, the *Essai sur le fluide électrique, considéré comme agent universel*: '*Mes yeux s'éteignent, mon esprit est las de travailler, c'est à vous à faire les plus grands efforts pur ne pas laisser échapper un fil qui peut vous conduire.*'[76] ('My eyes extinguish themselves, my spirit is weary of work, it is up to you to make greater efforts not to let slip a thread which may lead you on.')[77] It may have been the vistas he had glimpsed of a mass of yet undiscovered truths that drove Newton to make the famous remark near the end of his life: 'I don't know what I may seem to the world, but, as to myself, I seem to have been only like a boy playing on the sea shore, and diverting myself in now and then finding a smoother pebble or a prettier shell than ordinary, whilst the great ocean of truth lay all undiscovered before me.'[78]

Fortunately, the call was heeded, and electricity became a subject of intense investigation in the 18th century, often by capable amateurs. One such, a dyer from Canterbury, named Stephen Gray, made a breakthrough which established a new branch of the subject. His story is of special interest because of its connection with Newton and his followers. Though Gray had early on become a protégé of John Flamsteed, Newton gave him a position as assistant to Roger Cotes in setting up an observatory at Trinity College, Cambridge in 1707 and 1708. At the same time, he pursued his amateur interest in electricity. Unfortunately, a letter he sent to the Secretary of the Royal Society, Hans Sloane, ended up being refereed by Hauksbee, who,

---

[76]Tressan (1786, 1, xliv–xlv).

[77]Home (1982). A draft for the *Recensio Libri* (1722) (quoted Cohen, 1999, 282) makes essentially the same request.

[78]Spence (1820, 54). Spence attributed the anecdote to Andrew Michael Ramsay.

sensing the competition, simply incorporated what was new into his own work.[79]

When the observatory project failed, Gray had, for a time, to return to Canterbury, but around 1711 his health made it impossible for him to continue to work as a dyer, and Brook Taylor and John Keill, two of Newton's closest supporters tried to persuade him to become assistant demonstrator to the Royal Society. Though he refused the offer (possibly because Hauksbee was still demonstrator), a new opportunity arose in 1716 when another Newtonian disciple, John T. Desaguliers, who was now the Royal Society demonstrator, invited him to London as his assistant. Gray worked with Desaguliers for several years, living at his house, and a paper written in 1720, which was published in the Royal Society *Transactions* during Newton's Presidency, showed that such things as hair, silk and feathers, had the properties of recognised electrics such as amber. Gray proposed that electrical properties were a result of 'communication' of the electrical 'vertue', a *conduction* of electricity, and not something belonging to the actual electric.[80] Electricity was no longer purely static. In the same year, through the intervention of Sloane and Flamsteed, Gray became a pensioner at the Charterhouse and became free to pursue research as he wished. In February 1729, 2 years after Newton's death, he managed to transmit an electric current 52 feet along a silken thread, and then over hundreds of feet. Gray was elected a Fellow of the Royal Society and presented with its first Copley Medal.

Gray's discoveries directly inspired an explosion of electrical research during the rest of the century by Dufay, Franklin, Nollet, Priestley, Cavendish, Coulomb, Galvani, and many others, which culminated in Volta's discovery of the electric cell in 1800. Clearly, Newton and his supporters (with the possible exception of Hauksbee) had given encouragement to Gray over a considerable period, even though he had also had a long association with Flamsteed. Desaguliers subsequently created the terms 'conductor' and 'insulator' to refer to Gray's new discoveries, while Newton's friend, William Stukeley, described Gray as 'the father, at least first propagator, of electricity.'[81] Though a rather sensationalist account tried to argue otherwise, the evidence seems to suggest that Newton and the Newtonians positively welcomed such developments. Gray did not become a 'victim' of the long-running feud between Newton and Flamsteed, and electrical

---

[79] Gray (1708); Hauksbee (1708–1709).
[80] Gray (1720), *Philosophical Transactions*, vol. 31, 140–148.
[81] Stukeley (1882, 378).

science was able to make good progress, exactly as Newton would have wished.[82]

Eventually, by the end of the 18th century, it would be demonstrated that the static electric force was an inverse-square-law interaction, exactly like gravity, whose sources, called *charges*, were similar to gravitational mass, though being dissimilar in the fact that like charges repel while like masses attract, while magnetism would be similarly shown to require an inverse-square-law force between the poles, attractive for unlike and repulsive for like. From a fundamental point of view, once mass had been made a parameter on the same level as space and time, there was only charge to come. Space, time, mass and charge (when we extend charge to cover all the 'gauge' interactions) are the only known fundamental parameters in nature, and Newton's abstract representation of parameters is still the only way of approaching fundamentals that has been successful on a long-term basis. While such abstractness has not always been popular among physicists, quantities that are apparently unlike physically may have abstract similarities that are more fundamental, and no other method seems to have stood the test of time.[83]

Charge was a concept which evolved rather than being discovered, filling a conceptual gap needed to put electrical science into a mathematical form. Newton's discussion of electricity and magnetism as not being proportional to gross matter (or inertial mass) had suggested that some other thing must be responsible. Gravity is about mass, rather than matter, the 'quantity' of matter, rather than the kind of matter. Charge, unlike mass in principle, is about matter, rather than vacuum, the local rather than the non-local. Newton saw that all matter must be electrical to some *degree*, though the amount did not depend on the quantity of matter. If there were different degrees, then this implies some quantification, though such a concept was never stated explicitly. However, the idea that the attractive 'vertues' of electrical and magnetic bodies were found by integration over the elements of attractive 'vertue' of their component electrical and magnetic particles

---

[82] Clark and Clark (2000). Reviews, such as Silverman (2003) and the one by M. Peck in *Guidance*, 27, No. 4, 735, criticize the sensationalism of the book and the lack of credibility of the thesis. The title alone suggests popularist sensationalism in the line begun by Frank Manuel and continued by many ever since.

[83] Extensive treatment of these questions can be found in Rowlands (2007, 32–62). The fact that charge could be accommodated into dimensional analysis in the 19th century, originally without defining a new dimension, could be taken as an indication that the system of four parameters is closed. (There is a mathematical parallel with the closed system of quaternion units i, j, k presupposing the existence of 1.)

was the beginning of the idea of electric charge as an intrinsic property of matter.

Newton's intuition that the important force in the micro-realm was electricity would ultimately be proved correct. Every one of his positions has been vindicated, and his almost unified vision of the forces involved in electricity, magnetism and optics, and the cohesion of bodies, was ultimately achieved as 18th and 19th century scientists contributed to a gradual unification of electricity, magnetism, optics, chemistry, cohesion, capillarity, and almost every other non-gravitational phenomenon, as aspects of a single electromagnetic force. Only at the very end of the 19th century did the discovery of radioactivity by Becquerel lead to the discovery of phenomena, in particular, to the processes of α- and β-decay, which required the introduction of two entirely new forces: the strong and weak nuclear interactions. Today, the Standard Model of particle physics ascribes the entire operations of nature as due to the actions between particles of matter of only four fundamental forces, two of which are normally seated deep within material bodies. The massive study of electricity, starting in the 18th century and leading up to the present day, has also led to most of subsequent technology. Newton's breakthrough on the electric force was as important in principle as that on gravity, but its effect was nowhere near as significant because it came at the end of his life when he was in no position to carry it through to the same level of mathematical and experimental development, and his two most significant potential collaborators, Hauksbee and Cotes, both died young.

Chapter 8

Wave-particle Duality and the Unified Field

8.1. Conservation and Colour

From very early on in his optical work, Newton was faced with a dilemma. Two sets of experimental data, both originating in his own researches, seemed to require contradictory properties for light. The first involved the discovery of dispersion, the fact that a dense medium, such as glass, arranged in the form of a prism, was able to split apparently homogeneous white light into a spectrum, showing a continuous range of colours from violet to red, each of which had a different refractive index or degree of refrangibility (and a different velocity or momentum in a dispersive medium). However, the colour-forming property was not produced by the medium of the prism, but was conserved when the dispersed light was refracted through a second prism. The second discovery, of periodicity in the colours produced by thin films, seemed to suggest a wavelike attribute attached to the rays of light. Newton was unable to resolve his dilemma, but in leaving it unresolved he created the conditions for the modern doctrine of wave-particle duality in which the problem is essentially unresolvable.

In the first experiment, Newton's explanation of dispersion by a prism assumed that white light was a composite formed of rays of different refractive index; though he realised that the range of possible colours was infinite, he eventually identified seven principal ones on the analogy with the musical scale. The elongated spectrum that he observed for light from a circular hole passed through a prism was a result of the formation of a succession of overlapping circular images, each successive one being slightly displaced from the previous one as the prism refracted it at a new angle. With two prisms arranged crosswise, light was refracted on two planes at right angles. This showed that the shape of prism didn't stretch a spectrum at $45°$ to the axis of either prism in one direction more than another.[1]

Newton found that a slit worked better than a hole as light source, so initiating the use of this technique in spectroscopy. He also tried a triangular aperture with base a tenth of an inch wide and height an inch or two long, finding that on one side he obtained a pure spectrum and on the other an overlapping one, corresponding to the respective images of vertex and base. He further proposed one shaped like a parallelogram, of the same height, but of 0.05–0.1 inches width, or less.[2] His quantitative rule for establishing the purity of a spectrum was that 'the Mixture of the Rays in the refracted Spectrum . . . is to the Mixture in the direct and immediate Light of the Sun, as the breadth of the Spectrum is to the difference between the length and breadth of the same Spectrum.'[3]

Newton showed that, for a prism with refracting angle C, the minimum deviation or angle between incident and emergent rays (D), could be used to measure the refractive index according to the formula $n = sin\frac{1}{2}(D + C)/\sin\frac{1}{2}C$. Mainly using the experimental method described in the *Opticks*, it took a succession of amazingly powerful mathematicians in the mid-18th century, including Clairaut, d'Alembert, Bošković and Euler, to recover and extend Newton's mathematical derivation, even though the basic lemmas had been published in the *Optical Lectures* of 1728 and in Smith's *Compleat System of Opticks* 10 years later.[4] The same applied to the measurement of the chromatic dispersion or spectral length (Δn), where, using a convenient approximation, he was able to show that $\Delta n/n \approx \frac{1}{2}\Delta D$ cot i, for angle of incidence i. Again, a reconstruction of the equation could be made from Newton's experimental account in the *Opticks*.[5]

---

[1]Hall (1992, 103). *Lectiones Opticae*, Lecture 3, 32–36, OP I, 88–95; *Optica*, Lecture 2, II, 8–14, OP I, 440–449.

[2]*Opticks*, Book I, Part 1, Proposition 4. Shapiro (OP I, 460) reports Boursma (1971), 173–174, as showing 'that with these slits' and the lenses that he used, 'Newton would have attained nearly perfect spectral purity.'

[3]*Opticks*, Book I, Part 1, Proposition 4.

[4]See Shapiro, OP I, 178–180, *Lectiones opticae*, Lecture 9, 102, OP I, 176–180; *Optica*, I, Lecture 5, 29–31, OP I, 316–323; *Opticks*, Book I, Part I, Proposition 7; Clairaut (1762); d'Alembert (1764); Bošković (Boscovich) (1767); Euler (1762); Smith (1738), 'General remarks', §§391–398, 2: 66–67. Having assumed this in the earlier lecture, Newton finally showed that the minimum deviation required the rays to be refracted symmetrically on the two sides of the prism in *Lectiones opticae*, Lecture 18, Proposition 6, 186, OP I, 272–273, and *Optica*, I, Lecture 12, Proposition 25, 111, OP I, 394–395.

[5]Shapiro, OP I, 194–195; *Lectiones Opticae*, Lecture 10, 102–103, *OP* I, 190–195; *Opticks*, Book I, Part I, Proposition 7, Experiment 16. Newton calls the experiment 'nice and troublesome'.

Among a number of technical developments in the *Optical Lectures*, Newton devised a rule for finding the apparent magnitude of the image produced by a prism at minimum deviation, and showed, using an approximation, that, close to minimum deviation, the image would be at the same distance behind the prism as the object. However, though, at the time of the *Optical Lectures*, he believed that the minimum of dispersion coincided with the minimum deviation, he later realised that the image increased or decreased in size depending on which way the prism was rotated near minimum deviation. In the case of a dense medium becoming progressively denser, he used his own theory of maxima and minima to provide a 'neat' demonstration of a maximum for the dispersion and calculated the index of refraction at this maximum. Such a maximum, as Sergey Vavilov subsequently showed, requires the law of refraction to include separate terms depending on the index and the medium; this was true of both of Newton's erroneous laws of dispersion, as well as the later, more successful, Lorentz–Lorenz law. Then, in the last of the *Lectiones Opticae*, with the aid of a number of geometrical lemmas, Newton demonstrated that the angular dispersion increases with the angle of incidence whether the refraction takes place in a denser or rarer medium.[6]

Further work on the prismatic spectrum led Newton to believe that mirrors should be introduced into telescopes to avoid the problems of chromatic aberration in lenses — a phenomenon previously unknown which he himself first identified as a more significant problem of lenses than spherical aberration — and in 1668 he constructed a fully working model of a reflecting telescope based on this principle. This was a seriously difficult undertaking, and required many hours of patient labour, and extraordinary technical skill, especially to create a mirror of optical standard out of speculum metal. The design used a spherical primary mirror, with a plane secondary mirror placed diagonally close to its principal focus, which reflected the image through 90° to an eyepiece at the side of the telescope tube.

A second telescope, constructed in 1671, performed 'passably better'. Although only six inches long, at a 40 times magnification, it completely

---

[6] *Lectiones Opticae*, Lecture 18, Proposition 8, OP I, 276–279, *Optica*, Lecture 12, 115, OP I, 398–401. *Lectiones Opticae*, Lectures 14/15, 160–162, OP I, 240–245, *Optica*, I, Lecture 8, Proposition 11, 68–69, OP I, 356–359. See Shapiro, OP I, 276–277; *Opticks*, Book I, Part 1, Proposition 2, Experiment 3. *Lectiones Opticae*, Lectures 16/17, 174, OP I, 258–263; *Optica*, I, 99, OP I, 382–387; see Shapiro, OP I, 39 and 263, referring to Vavilov (1946). *Lectiones Opticae*, Lecture 18, 179–180, OP I, 268–269, *Optica*, I, Lecture 12, Proposition 20 and Scholium, 104–105, OP I, 392–393.

outperformed an equivalent 6-foot refractor, and was demonstrated to King Charles II in January 1672.[7] A third telescope, constructed with the help of Newton's friend John Wickins in 1671–1672, performed even better, and is now in the collection of the Royal Society in a restored form. The metal used in the primary mirror of this telescope was hardened by including arsenic rather than silver. The sensational success of the 1671 telescope at the Royal Society in 1672 no doubt contributed significantly to the fierce attack on Newton's work by Robert Hooke, who would have felt that optical and other instruments were his particular province. Conceivably, Newton's comment in his reply to Hooke that monochromatic light from his prisms produced superior microscopic images of insects and other natural objects may have been subtly aimed at an opponent, who, in his *Micrographia* of 1665, had shown himself the master of such studies.[8] After one or two attempts by the Royal Society to construct larger reflectors had failed through the inability of the manufacturers to make high enough quality mirrors, John Hadley, at a later date, constructed a 6-foot reflector to Newton's design, with a 600 times magnification, which was presented to the Royal Society after being demonstrated at the meeting of 12 January 1721.

The discovery of the composite nature of white light formed a major part of the *Optical Lectures* and the published *Opticks*, where we find statements such as: 'The Light of the Sun consists of Rays differently Refrangible.'[9] 'Refrangibility of the Rays of Light, is their Disposition to be refracted or turned out of their Way in passing out of one transparent Body or Medium into another.'[10] 'Lights which differ in Colour, differ also in Degrees of Refrangibility.'[11] '...I find that rays of equal incidence are refracted more and more after their disposition to exhibit colours in this order: red, yellow, green, blue and violet, with all their intermediate colours.'[12] 'The Sine of

---

[7]The first telescope was described in a 'Letter to a Friend', CUL, Add. 9597/2/18/3, 23 February 1669, NP; the second in 'An Accompt of a New Catadioptrical Telescope invented by Mr. Newton,' published in the *Philosophical Transactions of the Royal Society* on 25 March 1672, followed by two supplementary communications (NP).

[8]Newton to Oldenburg, 11 June 1672, *Corr.* I, 181, *Philosophical Transaction*, 7, 509, 1672; *Optica*, II, Lecture 12, 115, OP I, 559; *Opticks*, Book I, Part 1, Proposition 5.

[9]*Opticks*, Book I, Part 1, Proposition 2, Theorem 2.

[10]*Opticks*, Book I, Part 1, Definition 2.

[11]*Opticks*, Book I, Part 1, Proposition 1.

[12]Lohne (1965, 136); *Lectiones Opticae*, Lecture 3, 31, OP I, 88–89: 'at the same incidence rays that produce different colors undergo different refractions; namely, they are more and more refracted as they produce the successive colors in this order: red, yellow, green, blue, and purple, together with all their intermediate gradations.'

Incidence of every Ray considered apart, is to its Sine of Refraction in a given Ratio.'[13]

Reflection followed the same chromatic spectrum as refraction: 'Bodies reflect and refract Light by one and the same power, variously exercised in various Circumstances.'[14]

> The Rays of Light in going out of Glass into a *Vacuum*, are bent towards the Glass; and if they fall too obliquely on the *Vacuum*, they are bent backwards into the Glass, and totally reflected; and this Reflexion cannot be ascribed to the Resistance of an absolute *Vacuum*, but must be caused by the Power of the Glass attracting the Rays at their going out of it into the *Vacuum*, and bringing them back.[15]

There was a clear connection between the two phenomena, and also with diffraction or, as he termed it, 'inflection'. 'The Analogy between Reflexion and Refraction will appear by considering, that when Light passeth obliquely out of one Medium into another which refracts from the perpendicular, the greater is the difference of their refractive Density, the less Obliquity of Incidence is requisite to cause a total Reflexion.'[16] 'Do not the Rays which differ in Refrangibility differ also in Flexibility; and are they not by their different Inflexions separated from one another, so as after separation to make . . . Colours in . . . fringes. . .?'[17] 'Do not the Rays of Light which fall upon Bodies, and are reflected or refracted begin to bend before they arrive at the Bodies; and are they not reflected, refracted, and inflected, by one and the same Principle, acting variously in various Circumstances?'[18] '. . .the Rays which differ in Refrangibility may be parted and sorted from one another, . . .either by Refraction. . ., or by Reflexion. . .'[19] '. . .those Rays are more reflexible than others which are more refrangible.'[20]

Newton was fully aware that colour was a physiological response and not an intrinsic property of the light rays. He judged that the strongest response was to the rays producing yellow and orange; then those producing red and green; then blue; and finally indigo and violet.[21] In fact, the human eye has evolved to respond to the environment on Earth created by the peak in the

---

[13] *Opticks*, Book I, Part 1, Proposition 6, Theorem 5.
[14] *Opticks*, Book II, Part 3, Proposition 9.
[15] Q 29.
[16] *Opticks*, Book II, Part 3, Proposition 1.
[17] Q 2.
[18] Q 4.
[19] *Opticks*, Book I, Part 1, Proposition 2, Theorem 2.
[20] *Opticks*, Book I, Part 1, Proposition 3, Theorem 3.
[21] *Opticks*, Book I, Part 1, Proposition 7.

Sun's black body spectrum which is defined by its temperature, and so is particularly responsive to yellow.

Colours were either primary ('primitive', 'simple' or monochromatic) or 'compound'. Compound colours are formed at the borders of others in the spectrum, orange, for example, being formed at the border of the yellow and red regions, and capable of being formed by a compounding of red and yellow light. Intermediate colours are produced, in fact, by combining the ones near to them in the spectrum but not adjacent. So, while orange and green produce yellow, red and yellow produce orange, and violet and green produce blue.[22]

Newton devised a circular disc, with the seven colours, including indigo, arranged as sectors of the circle, though not all equally spaced; spinning the disc rapidly made it appear white. He realised that not all the colours were required to produce white, and hinted that this could be done with as few as three, and certainly with four or five: 'Whether it may be compounded of a mixture of three taken at equal distances in the circumference I do not know, but if four or five I do not much question but it may.'[23] However, two primary colours (adjacent ones on the disc) would not produce a sensation of white in this way. The closest Newton seems to have come to this was probably a near white or very pale green from mixing golden yellow and cyan. Though Huygens would claim, in correspondence, that white could be made out of yellow and blue (as, in his view, could all the other colours), it actually requires yellow and violet and would only give a true white with very pure colours. Newton replied that, though he had tried 'mixing all paires of uncompounded colours' and 'some were paler & nearer to white then others,' 'none could be truly called white.'[24]

Red and violet had to be primary as they were at the opposite ends of the spectrum and their compounding produced a non-spectral 'purple' (our magenta).[25] However, a combination of yellow and blue pigments did not produce the green of the spectrum, which occupied the central band, meaning that spectral green must also be primary or monochromatic, as also

---

[22] *Optica*, II, Lecture 8, 70, Proposition 4, OP I, 506–509.

[23] *Opticks*, Book I, Part 2, Proposition 6. The colours were arranged according to the musical scale, with the 'half-colours' orange and indigo fixed at the half-tone intervals between E and F and B and C.

[24] *Optica*, II, Lecture 8, Proposition 4, 70, OP I, 506–507; Oldenburg to Newton, 18 January 1673, *Corr.* I, 255–256, *Philosophical Transactions*, 8, 6086, 1673; Newton to Oldenburg, 3 April 1673, *Corr.* I, 265, *Philosophical Transactions*, 8, 6110, 1673. See Shapiro, OP I, 507.

[25] *Opticks*, Book I, Part 2, Proposition 6.

appeared to be the case from Newton's experiments on boundary colours.[26] Although yellow light projected onto a blue body made it appear green, this was not true colour mixing, but a consequence of the fact that more of the green-producing rays in the yellow light were reflected than were the yellow-producing ones.[27]

Though Newton tended to follow the prevailing opinion that mixing coloured lights and coloured pigments or powders gave the same results, and he maintained that the individual powder grains in the mixed powders, examined under microscopes, still retained their original colours, he also realised that the colours of powders were frequently impure, and were really mixtures of many colours with only the dominant one visible. A powder that appeared, for example, blue could contain cyan, indigo, green and even yellow, as well as violet.[28] In some cases, as with plates of coloured glass, he observed that the colour mixing could be subtractive rather than additive.[29]

From his very first researches, Newton had been able to explain boundary colours as being produced by the colours of light, separated by unequal refrangibility, that fall on the boundary and reach the eye, depending on the angle of viewing. Many contemporaries, including Charleton, Hooke and Pardies, and the poet Johann Wolfgang von Goethe long afterwards, were inclined to believe that colour was formed only on the boundary of light and shadow, but Newton refuted this decisively in a number of experiments.[30] In the optical lectures, he described using a wide, rectangular slit formed by two straight-edged obstacles. Initially, he saw violet and blue at one edge and yellow and red at the other, with white light in between. As he narrowed the slit width, he saw the colours on each side begin to coalesce. Narrowing it further, he saw green appear for the first time and then a complete spectrum.[31] His work also showed that white, black and grey were not 'colours', as was generally thought at the time, but merely different

---

[26] Newton to Oldenburg, 11 June 1672, *Corr.* I, 181. *Lectiones Opticae*, Lecture 7, 84, OP I, 150–151. *Optica*, II, Lecture 10, 84, OP I, 526–527.

[27] *Optica*, II, Lecture 9, 72, Proposition 4, OP I, 510–511.

[28] *Lectiones Opticae*, Lecture 4, 51, OP, I, 108–113; *Optica*, II, Lecture 5, 38, OP I, 472–475, Lecture 9, Proposition 4, 70, OP I, 506–509.

[29] *Optica*, II, Lecture 9, 75–76, OP I, 514–516, 519.

[30] *Optica*, II, Lecture 12, 103–106, OP I, 548–553; Goethe (1791–1792).

[31] *Lectiones Opticae*, Lecture 7, 84, OP I, 150–151, *Optica*, II, Lecture 10, 84, OP I, 526–529. Shapiro describes this as 'explicat[ing] the transition from the polychromatic boundary colors to the monochromatic spectral colors' (OP I, 33).

degrees of brightness in white light, with black representing the absence of all colour, the extreme case of subtractive mixing using painters' colours.[32]

A number of substances, including lignum nephriticum, a wood originating in Mexico which was used in water infusion for medical purposes, and also gold leaf and coloured glass, had the properties of transmitting one colour and reflecting another. This could now be explained:

> The odd phenomena of an infusion of lignum nephriticum, leaf gold, fragments of coloured glass, and some other transparently coloured bodies, appearing in one position of one colour, and of another in another, are on these grounds no longer riddles. For, those are substances apt to reflect one sort of light and transmit another; as may be seen in a dark room, by illuminating them with similar or uncompounded light. For then they appear of that colour only, with which they are illuminated, but yet in one position more vivid and luminous than another accordingly as thay are disposed more or less to reflect or transmit the incident colour.[33]

'Lignum nephriticum,' he wrote, 'sliced and about a handful infused, in three or four pints of fair water, for a night, the liquor (looked on, in a clear vial) reflects blue rays; and transmits yellow ones. And if the liquor, being too much impregnated, appears, (when looked through,) of a dark red, it may be diluted with fair water, till it appears of a golden colour.'[34] Transmission and reflection were separate processes, which, in such cases, and seemingly in the case of the stained glass found in churches, produced light with complementary colours, though the absolute generality of the rule was broken by the existence of glass which transmitted and reflected light of the *same* colour.[35]

Newton's experiments showed that coloured rays were *not* produced by alterations of white light within the prism but were present within the white light waiting to be dispersed when an object like a prism split them up according to their refractive indices.[36] They could be seen even

---

[32] *Lectiones Opticae*, Lecture 4, 51, OP I, 108–113; Newton to Oldenburg (reply to Pardies), 13 April 1672, *Philosophical Transactions*, 7, 4092–4093, *Corr*. I, 141–142; Newton to Oldenburg (reply to Hooke), 11 June 1672, *Philosophical Transactions*, 7, 5099–5010, *Corr*. I, 183–184; *Opticks*, Book I, Part II, Proposition 5, Experiment 15, 150–154. *Optica*, II, Lecture 8, 69, OP I, 504–507 (on black).

[33] Newton to Oldenburg, 6 February 1672, *Corr*. I, 99.

[34] *De coloribus*, 2; quoted *Corr*. I, 190.

[35] *Of Colours*; *Optica*, II, Lecture 9, 73–74, OP I, 510–515, *A New Theory*, Proposition 11, Newton to Oldenburg, 6 February 1672, *Corr*. I, 99; see Shapiro, OP I, 514–515. Selective absorption was another process which complicated the picture.

[36] Newton to Oldenburg, 6 February 1672, *Corr*. I, 97; draft Q 21/29 for *Opticks*, 1706, CUL, Add f. 291^r; Bechler (1973, 33).

when rays passed through parallel refracting surfaces as in a glass block, a parallelepiped made of solid glass or a hollow one filled with water, or a pair of prisms, one inverted and one upright, separated but with their bases parallel, though generally only at distances close to the final refracting surface.[37]

The experiments, including the one that showed that individual coloured rays were unchanged in passing through a second prism (the *experimentum crucis*, as he called it), convinced Newton in his belief that rays of light were physical objects and not just geometrical constructions. That rays were physical things was an ancient tradition, stemming at least from the time of Euclid's description of visual perception in terms of countable rays emitted by the eye.[38] Hobbes had defined the ray as the path by which the 'motion from the luminous body' was transmitted through a medium, but he still felt that rays had physical characteristics such as a finite volume.[39] One possibility that was available was that the rays were small bodies or corpuscles, atoms of light, as it were. Atomism was a well-discussed topic in the 17th century, and many of Newton's contemporaries, the proponents of the 'mechanical philosophy' were atomists of one sort or another. One of the sources for Newton's early notebook entries as a student was the popular scientific writer, Charleton, who had developed an atomic theory, incorporating particles of light, partly on the basis of Gassendi's revival of the ancient atomism of Epicurus and Lucretius, and partly on misreadings of the mechanical analogies in the works of Descartes.[40]

The possibility was very much in mind when Newton wrote his first paper on optics for transmission to the Royal Society. He did not, however, structure his paper on this hypothesis, as he was sure that he had a result which transcended all hypotheses. Nevertheless, after testing the various alternative possibilities which suggested themselves before he performed his *Experimentum Crucis*, and gradually eliminating them, Newton remarked that 'if the Rays of light should possibly be globular bodies and by their oblique passage out of one medium into another acquire a circulating motion, they ought to feel the greater resistance from the ambient Æther, on that side, where the motions conspire, and thence be continually bowed to the other.'[41] The inclusion of a mechanistic corpuscular hypothesis — even

[37] *Optica*, II, Lecture 13, 118–123, OP I, 563–571.
[38] Euclid, *Optiks*, Proposition 3.
[39] Hobbes (1644, 576–587); Shapiro (1974, 148).
[40] Charleton (1654).
[41] *A New Theory*, Corr. I, 97.

though it was incorporated only as illustration of an unsuccessful alternative to the main idea — gave the opportunity for those who had yet other hypotheses to assert the superiority of their own, with disastrous consequences for the immediate acceptance of Newton's ideas.

When the climate at the Royal Society had changed for the better, in 1675, he wrote about the possibility of corpuscles in a work specifically designated by the title *Hypothesis of Light*, supposing that Light was neither 'Æther nor its vibrating motion,' but something propagated from 'shining bodies', which one could *imagine* as a stream of very small and swift 'Corpuscles', driven 'forward by a Principle of motion.'[42] A similar idea was later presented in the *Opticks* in the section which was specifically labelled 'Queries' in order to allow the discussion of hypothetical ideas.

Nevertheless, though Newton always avoided any presentation of a corpuscular model of light as anything but a plausible hypothesis, it was one that underpinned the whole of his theoretical reasoning on the nature of light. Whether or not light could be described in terms of mechanistic corpuscles, he always insisted that it had to have particle-*type* properties because colours of rays were immutable in optical processes — exactly the same point as was later made by the quantum theorists of the atom.[43]

Fundamental to Newton's theorising on optics was his belief that light was subject, in some sense, to the abstract property of *conservation*. His novel theory of chromatic dispersion was conditioned by his observation that individual light rays remained unchanged during reflection or refraction, and so something had to be conserved in optical processes. The theory, although implying the existence of some kind of conserving, and therefore particle-like, mass-related property within light, was not, in any sense, based on a mechanical model, but on the fulfilment of an abstract condition. The distinction was deep and never appreciated by his contemporaries, who were 'mechanical philosophers' without any conception of an alternative.

If light is made of very small bodies, something will certainly be conserved. However, as Newton knew from an early stage in his work, the bodies, if they existed, were not exactly identical to those of ordinary matter. Reflection and refraction, for example, were not due to mechanical impact, '...but some other power by which those solid parts act on Light at a distance.'[44] '...& the refraction I conceive to proceed from ye continuall

---

[42] *Hyp., Corr.* I, 370.
[43] *A New Theory, Corr.* I, 92–102.
[44] *Opticks*, Book II, Part 3, Proposition 14.

incurvation of the ray all the while it is passing the Physical Superficies.'[45]
In the early *Hypothesis of Light*, he argued that surfaces could not be made
of particles all pointing the same way:

> In polishing glasse or metall it is not to be imagined that Sand, Putty or
> other fretting pouders, should wear the surface so regularly as to make the
> front of every particle exactly plaine, and all those plaines look the same
> way, as they ought to do in well polished bodyes, were reflexion performed
> by their parts: but that those fretting pouders should wear the bodies first to
> a coarse ruggednes, such as is sensible, and then to a finer & finer ruggednes,
> till it be so fine that the Æthereall Superficies eavenly overspreads it, and
> so makes the body put on the appearance of a polish, it is a very naturall &
> intelligible supposition. So in fluids it is not well to be conceived, that the
> Surfaces of their parts should be all plaine, & the plains of the Superficiall
> parts always kept looking all the same way, notwithstanding that they are
> in perpetuall motion...[46]

The conservation property, though fundamental to Newtonian thinking,
was, perhaps because of its abstractness, a problem with contemporaries,
such as Leibniz, who had conservation properties of their own. Leibniz, as
we have seen, thought that the idea of 'primitive bodies which always kept
their colours' was not impossible, but he couldn't see how it could explain
the laws of refraction.[47]

8.2. Particles with Periodicity

Newton's second major optical experiment, on the coloured fringes generated
in thin films, demonstrated that light was not only a stream of particles.
There was a periodicity which could be observed and even quantitatively
measured using what we would now describe as interference phenomena in
thin films. Newton eventually convinced himself after further experiments,
involving thick films, that periodicity was almost an *innate* property of light,
with a quantitative measure that would become known in his mature work
as the 'interval of fits': 'Suppose the Light reflected most copiously at these
thicknesses be the bright citrine yellow, or confine of yellow and orange...'[48]
'If the Rays which paint the Colour in the Confine of yellow and orange pass
perpendicularly out of any Medium into Air, the Intervals of their Fits of
easy Reflexion are the 1/89000th part of an inch. And of the same length are

---

[45] *Hyp., Corr.* I, 371.
[46] *Hyp., Corr.* I, 375.
[47] Leibniz to Huygens, 26 April 1694; Brewster (1855, 1: 149–150).
[48] *Opticks*, Book II, Part 2.

the Intervals of their Fits of easy Transmission.'[49] Ultimately, Newton's work led to the science of interferometry and a very precise method of measuring lengths.

Newton's classic understanding of periodicity, in his more hypothetical earlier work, was to see it as the action of light rays on the medium.

> ...if light be incident on a thin Skin or plate of any transparent body, the waves excited by its passage through the first Superficies, overtakeing it one after another, till it arrive at the second Superficies, will cause it to be there reflected or refracted accordingly as the condensed or expanded part of the wave overtakes it there. If the plate be of such a thicknesse, that the condensed part of the first wave overtake the ray at the second Superficies, it must be reflected there; if double that thicknesse that the following rarefied part of the wave, that is, the space between that and the next wave, overtake it, *there* it must be transmitted; if triple the thicknesse yt the condensed part of the second wave overtake it, *there* it must be reflected, ... & so on....[50]
>
> And for explaining this, I suppose, that the rays when they impinge on the rigid resisting æthereall Superficies, as they are acted upon by it, so they react upon it and cause vibrations in it, as stones throwne into water do in its Surface; and that these vibrations are propagated every way into both the rarer & denser Mediums, as the vibrations of Air wch cause Sound are from a Stroke, but yet continue Strongest where they began, & alternately contract & dilate the æther in that Physicall Superficies.[51]

However, his characteristic stance, especially in later work, was to adopt the fewest possible assumptions compatible with the mathematical description of a phenomenon. This was why the more descriptive models of the earlier work were reduced to the almost model-free concept of fits of easy reflection and easy refraction. They could, in particular, be applied to his idea that the particlelike 'rays' did not reflect or refract by mechanical collisions with the particles of a reflecting or refracting surface.[52] In adopting fits he had taken up the most positivist stance which he found compatible with the experimental evidence.[53] He had reduced the problem as far as possible to a statement of the phenomena requiring no hypothetical explanation; but the theory, at least, suggests that he came to view some part of the wave aspect as being intrinsic to light itself, and not just something produced by the interaction of light with media, whether material or aetherial.

---

[49] *Opticks*, Book II, Part 3, Proposition 18.
[50] *Hyp., Corr.* I, 378.
[51] *Hyp., Corr.* I, 374.
[52] *De vi Electrica, Corr.* V, 365.
[53] *Opticks*, Book II, Part 3, Proposition 12.

This view was confirmed by his later work on the interference rings produced in thick polished plates, where he found experimental results, involving successions of more than thirty thousand alternations of fits of reflection and refraction, agreed with calculations based on the theory he had already developed for thin films: 'And thus I satisfy'd my self, that these Rings were of the same kind and Original with those of the Plates, and by consequence that the Fits or alternate Dispositions of the Rays to be reflected and transmitted are propagated to great distances from every reflecting and refracting Surface.'[54] So, after much thought, he began to consider that even the fits had to be associated with the immutable nature of light corpuscles:

> from...Observation, the Light reflected by thin Plates of Air and Glass, which to the naked Eye appear'd evenly white all over the Plate, did through a Prism appear waved with many Successions of Light and Darkness made by alternate Fits of easy Reflexion and easy Transmission, the Prism severing and distinguishing the Waves of which the white reflected Light was composed.... And hence Light is in Fits of easy Reflexion and easy Transmission, before its Incidence on transparent Bodies. And probably it is put into such fits at its first emission from luminous Bodies, and continues in them during all its progress.[55]

If Newton's first instinct was to believe that the wavelike periodicity produced by light rays was extrinsic to them, his inability to explain away the periodicity of both thick and thin film effects forced him eventually to accept a theory in which wavelike properties might also be innate.

As we have seen, his early optical work had already demonstrated that the property by which light produced the sensation of a particular colour was unchangeable and not subject to modification during reflection or refraction, and this, he was certain, must indicate a particulate nature for light. Since periodicity was directly associated with the production of colour in Newton's experiments, this meant that some kind of periodicity or wavelike behaviour had to be associated with the actual particles of which light was composed. Newton did not eliminate this extraordinary ambiguity from his work precisely because *it was present in the data itself*; no other 17th century scientist showed such an uncompromising attitude to the primacy of experimental evidence. The *Opticks* supplies his mature position that the actual nature of the cause of periodicity need not be specified. We

---

[54] *Opticks*, Book II, Part 4, Observation 8.
[55] *Opticks*, Book II, Part 3, Proposition 13.

need not inquire whether it is a circulating or vibrating motion of the ray, or of the medium, or something else.[56]

There were two reasons why, for all his recognition of periodicity, Newton never adopted a complete wave theory of light of the kind favoured by Hooke or Huygens, though his concept of interval of fits can be seen as equivalent to wavelength and he applied it correctly to the sequence of colours. One was his belief that light was in some way primarily particulate because it had immutable properties, the other was that the wave theory did not explain why light should travel in straight lines, and not spread out everywhere into the surrounding medium. The standard wave theory could not explain why light had a forward momentum. For Newton, if light travelled in straight lines it must be a stream of particles not merely a wave pulse in a medium; something must direct the wave forward. But the particles of light must somehow produce vibrations, and these vibrations must be characteristic of the particular type of light with which they are associated. The analogy with water waves produced by a stone made it possible to pursue this idea. Light could not be like the vibrations produced by a single stone thrown into water but it could be like the vibrations made by a long series of stones.

He also considered that, if light and matter interact, and can be transformed into one another, then they must both be part of the same scheme of nature. One cannot be particulate if the other is not. There cannot be two fundamentally different types of substance. Alternative schemes, such as a pressure wave with no forward motion or motion transmitted to all parts instantaneously without finite speed, would fail for other reasons. As he says in a draft of Query 20/28:

> The common opinions are that light is made either by pression or motion propagated through a fluid or by bodies projected. If it consisted only in pression propagated without actual motion, it would not be able to agitate & heat the bodies which refract & reflect it. If it consisted in motion propagated to all distances in an instant, it would require an infinite force to generate that motion.[57]

Newton never deviated from his stance that a particle-related property was a fundamental aspect of the nature of light, but he did not put forward a mechanistic view of light as composed of material-like particles, as was long believed. The subtlety of his views was lost on his 18th century successors, and it was largely their view of his corpuscles as rigid mechanical bodies

---

[56] *Opticks*, Book II, Part 3, Proposition 13.

[57] Draft of Q 20/28, UL Cambridge, MS Add. 3970.3, 287^r, NP.

which subsequently prevailed. When the corpuscular theory promoted by physicists like Laplace and Malus was overtaken by the wave theory of Young and Fresnel, Newton's *Opticks* fell into such neglect that it was not reprinted again until 1931. As part of their propaganda effort, the 19th century wave theorists generated a myth of Newton's dogmatism over the corpuscular theory of light, and it proved too convenient a story to be challenged for more than a century. Newton, however, had never made 'an explicit public commitment to the corpuscular theory,' and most 18th century corpuscular theorists were neither specifically Newtonian nor direct followers of Newton himself. The evidence simply doesn't exist that would 'show that [Newton's] authority was responsible for the popularity of his views,' which were, in any case, by no means universally admired.[58]

Because periodicity was an essential property of light, Newton produced many models of wave theory over the years, trying to reconcile the straight line motion and presumed particle nature with the obvious periodicity shown by thin film effects, but he was always consistent in his approach and cannot be said to have wavered between particle and wave theories. He never accepted a theory which did not give light a forward momentum and rejected all explanations unless they incorporated a periodicity characteristic of the waves themselves. The investigator was free to assume any hypothesis in which these components were accepted and, in the present state of knowledge, none of these was preferable to any other; even the idea of fits, he said, had no better grounds for being accepted than any of the other wave-type ideas he had postulated.

8.3. A Dualistic Theory

While the long controversy between the wave and particle theories is now resolved with a dualistic explanation, in which all optical effects may be described either in particulate or wavelike terms, not all such descriptions have as yet been discovered. The abstract duality allows a wide range of possibility for 'physical' realisation and the evidence presented by the versions known suggests that they are probably not exclusive. For example, though, at the most fundamental level, the quantum explanation has often been assumed to require particulate features, a Maxwellian wavelike explanation of vacuum also exists, based on the zero-point energy. Like the many possible 'interpretations' of quantum mechanics, such alternative physical explanations help to establish that no single interpretation is more

---

[58]Cantor (1983, 11, 200).

valid than any other. Newton's varied writings on light suggest that he came to believe in something which combined aspects of both particle and wave theories, not exactly, to be sure, as we do today, but not entirely dissimilar.

There was a dualism in Newton's work right from the beginning. The manuscript of his student notebook, *Quaestiones Quaedam Philosophicae*, has a drawing of a 'Globulus of light' surrounded by 'a cone of subtile matter wch it carrys before it the better to cut ye ether, wch serves also to reflect it from other bodys.'[59] The concept stemmed from his early reading in Descartes, though as always there was an additional subtly Newtonian interpretation in which the particle aspect is transferred from the aether to the body of light. In a later reply to criticisms by Pardies of his own work, he wrote concerning dispersion, 'what can be imagined from the Cartesian hypothesis — a like diffusion of effort or pressure of the globules, just as is supposed in accounting for the tail of a comet,' and considered the explanation to be just as valid as Grimaldi's 'substances in rapid motion' and Hooke's use of wave motion.[60]

The key experiments, done soon after the notebook entry, made the duality an inevitable consequence of approaching the data analytically. Isaac Barrow took a pragmatic (and, as it happens, prophetic) approach in his *Optical Lectures* of 1668, writing: 'I assume that light is produced both ways, both through corporeal effluvia; and through a continuous impulse; and it will be better to attribute some of its effects to the latter, others to the former.'[61] However, Barrow's duality is between corpuscles and 'a continuous impulse', rather than corpuscles and waves or periodic phenomena, and Newton, who already had the basic ideas of conservation of colour and periodicity, is unlikely to have been strongly influenced by his mentor in this area.

Newton's dilemma was that the conflicting requirements of innate particle-type properties, such as a conserved mass, at the same time as wavelike periodicity, required light particles and aether vibrations to have, both different speeds, and speeds which responded in a different way in refracting and dispersive media. This is the key point in modern wave-particle duality. Newton was the first of many to confront it and the first to

---

[59]QQP; 'Of species visible', f. 104^v, NP.

[60]Newton to Oldenburg, reply to letter from Pardies, 10 June 1672: *Corr.* I, 163–168, paraphrase 168–171, 169. The Latin text referring to Grimaldi reads '*puta ex Hypothesi P. Grimaldi, per diffusionem luminis, quod supponitur esse substantia quædam rapidissimè mota*'; Mr. *Newtons* Answer, NP.

[61]Barrow (1669, I. 6); *Math. Works*, 15.

attempt a resolution. The resolution required an approach to physics which could not be described as mechanistic, and, through the seemingly arbitrary concept of fits, this non-mechanistic element entered the mainstream of physics, never entirely to be lost from it. The originality of Newton's attempt to overcome the limitations of the current mechanistic philosophy as practised with success by contemporaries like Hooke and Huygens has never been fully appreciated. Later authors have tended to stress his failure to come to terms with the facts according to his supposed canons of scientific procedure, and have overlooked the truly radical nature of what he attempted.

It is possible, in fact, to see in Newton's own work a struggle to *find* a particle theory which would fit the experimental evidence rather than a desire to work out the consequences of some predetermined mechanistic view of nature, and to produce, as a result of that struggle, a particle theory of light which is much closer to the modern quantum theory than is usually supposed. None of the 17th century theories, wave or particle, gave a complete explanation of the properties of light, even as then known, but Newton's was the only theory which combined both wave and particle aspects. His approach to particle theory was not mechanistic but *minimalistic*. His was a zero option theory, intended to explain the fact that the fundamental characteristics of light rays were not changed during optical processes, but at the same time there were also strongly wave-type aspects, in particular the concept of mathematical periodicity which he discovered when he investigated interference rings and which he explained in his theory of fits.

Though Newton insisted that the corpuscularity of light was necessary to explain the unchangeable colours of light rays, in the end, after the failure of numerous specific models, it was only the *idea* of corpuscularity which was retained, and even then it had to be combined with an equally abstract idea of wavelike periodicity. It is only today, after the development of quantum mechanics, that such dualistic notions can be fully appreciated. As I. Bernard Cohen has pointed out, Newton, by adopting dualistic views, had 'violated one of the major canons of 19th-century physics, which held that whenever there are two conflicting theories, a crucial experiment must always decide uniquely in favour of one or the other,'[62] though it could be argued that he only had himself to blame for stressing, for his immediate purposes in 1672, the exceptional importance of an *experimentum crucis*. For the 19th-century physicist, it was not only that Foucault's experiment had shown that Newton

---

[62] Cohen (1952, viii).

had backed the losing argument on the crucial question; it was also that he had used an unacceptable scientific style — one that lay quite outside the mathematico-deductive tradition established at the beginning of that century.

Of course, as Cohen has also stressed, Newton's simultaneous use of apparently incompatible ideas is not as puzzling to us as it was to our 19th-century predecessors. But there is still no consensus as to why this remarkable 'coincidence' in viewpoint should have occurred. It would seem, however, that the style of progressive abstraction which was natural to Newton has been gradually forced upon the whole of physics as a result of the actual discoveries made over a long period, and that it was always the style most likely to lead to the most fundamental aspects of knowledge. Sir Edmund Whittaker's Introduction to the edition of the *Opticks* of 1931, spoke of the 'considerable analogies' that were to be observed between Newton's ideas and 'modern views', and quoted a 'distinguished physicist' as saying of the theory of fits: 'After being regarded for a generation as an artificial attempt to save a dying theory, we have proved this guess of Newton's to be a supreme example of the intuition of genius.'[63] It was not, however, a 'guess'; it was, in fact, an example of Newton's standard method of analytical abstraction.

None of his statements about light were the speculative guesses they have sometimes been assumed to be. Where we can trace the origin of the idea, it is always based on the process of reducing assumptions. After long consideration of the evidence, Newton adopts, as we have seen, what may be called a minimalist or zero-option position, even against his own prejudices. Thus, his celebrated anticipation of wave-particle duality is not an amazing coincidence, as has often been claimed, but a realisation that light has fundamentally contradictory properties. Newton was even more correct in effectively coming to the position that light is neither waves nor particles, but a concept that requires a more fundamental and abstract explanation.

Newton's anticipation of 20th century ideas, then, was not the result of some inexplicable 'coincidence' in viewpoint, but the consequence of his taking almost the same positivistic attitude to the same experimental evidence; the ambiguities which he experienced are the same as the ambiguities which we have been obliged to recognise as unavoidable because the experimental evidence which led to them has remained unchanged to this day, and Newton's presentation was almost devoid of the kind of

---

[63]Whittaker (1931/1952, lxiii).

theoretical content which would have limited its scope. Consequently, his light corpuscles, though defined by their mass and momentum, were nothing like the particles of mechanistic philosophies and even their particulate nature was nothing but a necessary consequence of certain experimental results and not due to the assumption of any particular hypothetical model.

The innovation in scientific methodology, which this required, has a great deal of relevance for fundamental physics in the 21st century, though its importance to the study of the ultimate nature of light has gone unrecognised. The key question 'Why was Newton a dualist?' has not been satisfactorily addressed. Otherwise, we would have become aware that Newton's employment of wave-particle duality to explain light was not a curious and accidental anticipation of modern views, as is often supposed, but a result of his discovery of the most profound and effective system of scientific methodology, the gradual rediscovery of which has led to the most important and 'revolutionary' advances of subsequent times.

8.4. The Early Unified Theory

Newton knew that his work in optics was incomplete, and that the basic first principles underlying Nature had yet to be discovered. As he wrote in a draft Query:

> Much remains to be discovered concerning the nature of Light, much more concerning the nature of fire & other bodies which emit reflect, refract inflect & stifle it & concerning the heat, motion & powers by which it is emitted reflected refracted inflected & stifled. Many experiments are wanting for completing the Analysis of this part of Nature & coming to a clear & distinct knowledge of all the causes of these things, many more for perfecting the Analysis of all Nature & making a full & clear discovery of all the first Principles of Natural Philosophy. To compass this is a work which requires many heads & hands & a long time & yet this ought to be done before we proceed from the first Principles by Composition to explain {all}Nature.[64]

He twice, however, attempted to combine the forces of nature in what may be described as a 'unified field' theory. The first was an early attempt at a 'theory of everything'.[65] At the time he was a young man, full of confidence that such a thing was possible, though fully aware that the particular hypothesis he was putting forward was highly tentative. The second was a much

---

[64]Draft Q 25, Add. 3970.3, CUL, f. 244^v, NP.

[65]*Hyp., Corr.* I, 361–386. William R. Newman uses 'theory of everything' to refer to the related but earlier and more explicitly alchemical *Of Natures Obvious Laws*, c 1672, NPA.

more subdued attempt, in old age, to make sense of the proliferation of natural forces with which he was confronted. Both required the idea of an aether.

Assessment of these ideas has been made more difficult by modern prejudices against the aether based on its supposed elimination by Einstein's special theory of relativity. Essentially, if Newton finds a mode of doing physics that dispenses with the aether using action at a distance between point particles, as he does at the time of the *Principia* (1687), then this is line with current modes of thought and so is to be commended. If he feels he has to revert to aethereal explanations, as at the time of the third edition of the *Opticks* (1717), then this is a retrograde step taken by an old man, retreating to a safer mode of thinking, and so to be considered as unfortunate.

This is, in fact, a total misunderstanding both of Newton's scientific development and of the meaning of the aether concept in physics.[66] Newton was interested in aethereal ideas throughout his entire career, but they were not a means of replacing action at a distance, or vice versa. They were alternative ways of looking at the same physical information. From the modern point of view, although we no longer use the word 'aether', we still use same kind of idea when we talk about fields and 'vacuum'. There are always two ways of defining a physical system and its relationship with the rest of the universe — either by starting with the system and working out its connections with the universe through repeated iteration, or by attempting to define the rest of the universe and working out how the system fits into it. Quantum mechanics allows us to do this in a very specific way by defining fundamental particles as excitations in a quantum field, and by opposing particles and vacuum, the local and the non-local. Ultimately, the duality can be seen as originating in a fundamental one in information processing between the iterative and the recursive viewpoints.

The duality between particle-based and aethereal explanations is present in Newton's work at nearly all periods, though he sometimes emphasises one interpretation and sometimes the other. A mathematically-based calculation will tend to be thought out in terms of point-particles; an attempt at finding a physical 'meaning' may involve a discussion of possible aethers. There is no necessary contradiction between these viewpoints — they are looking at the same information from different perspectives. However, the mathematically-based calculations always took precedence over any attempted 'physical' explanation, as the latter were always considered speculative, and there seems to have been a period, around the time of the first edition of the

---

[66] Hall (1992, 357) goes so far as to make Newton a 'progenitor of the great aether delusion,' but, if Newton is deluded, then so are all modern quantum field theorists!

Principia, when physical explanations were avoided altogether. Newton was always aware that the mathematical calculations would stand irrespective of the status of the physical explanation.

At an early stage in his career, Newton was influenced, like many others, by the speculations of Descartes, and his explanations of the motions of planets as being due to the actions of vortices in a dense material aether. It took a long time to completely eradicate this notion from his thinking, with the physical side of his analysis being completely outpaced by the mathematical. The attraction of the aether theory was that it offered the possibility of a complete explanation of all the phenomena of nature in a great synthesis, a 'theory of everything'. The *Hypothesis of Light*, of 1675, was just such a synthesis, combining the early aether ideas with the more abstract and analytical way of thinking developed to explain experimental results in optics, and incorporating proto-chemical ideas from his parallel investigations in alchemy.[67] There was even, seemingly, an attempt to reconcile his developing idea of an inverse square law with the aethereal motions which came from the other sources.

The *Hypothesis* certainly had an alchemical background. Early chemical manuscripts seem to suggest that chemical processes require the action of a universal vital agent, which, more than the Cartesian aether, seems to be the unifying principle that Newton believed must exist in nature. A list of prepositions, from about 1669, says that:

> The vital agent diffused through everything in the earth is one and the same. And it is a mercurial spirit extremely subtle and supremely volatile, which is dispersed through every place. The general method of operation of this agent is the same in all things; that is, it is excited to action by a gentle heat, but driven away by a great one, and when it is introduced into a mass of substances its first action is to putrefy and confound into chaos; then it proceeds to generation.[68]

Another alchemical manuscript, of the 1670s, *Of Natures Obvious Laws & Processes in Vegetation* (referred to by Westfall as *The Vegetation of Metals*), seems like a precursor to the *Hypothesis*:

> This is the subtil spirit which searches the most hiden recesses of all grosser matter which enters their smallest pores & divides them more subtly then any other materiall power what ever this is Natures universall agent, her secret fire, the onely ferment & principle of all vegetation. The material soule of all matter which being constantly inspired from above

---

[67] *Hyp.*, *Corr.* I, 361–386.
[68] List of prepositions, c 1669, KCC, Keynes MS, 12; Westfall (1980, 304–305).

pervades & concretes with it into one form & then if incited by a gentle heat actuates & enlivens it... Note that tis more probable the aether is but a vehicle to some more active spirit.... This spirit perhaps is the body of light because both have a prodigious active principle both are perpetuall workers.... Noe substance so indifferently, subtily & swiftly pervades all things as light & noe spirit searches bodies so subtily percingly & quickly as the vegetable spirit.[69]

By 1675, Newton had also experimented on electricity, and imagined something like a vital agent becoming manifest. Still influenced by the Cartesian aether, or at least the gravitational aether he had written about in his early notebook, Newton saw that a universal aether of some kind might be able to unify everything in the physical world — gravity, capillary action, light, electricity and magnetism — and be interconvertible with matter, condensed into material form 'as it were by precipitation,' though this medium might be 'compounded partly of the maine flegmatic body of æther partly of other various æthereall Spirits, much after the manner that Air is compounded of the flegmatic body of Air intermixt with various vapours & exhalations. For the Electric & Magnetic effluvia and gravitating principle seem to argue such variety.'[70]

The exposition which followed fused alchemy, the mechanical philosophy and Newton's developing abstract style of reasoning in a way that he never repeated in his work. Maybe aether was the origin of all matter.

Perhaps the whole frame of Nature may be nothing but various Contextures of some certaine aethereall Spirits or vapours condens'd as it were by prae-cipitation, much after the manner that vapours are condensed into water or exhalations into grosser Substances, though not so easily condensible; and after condensation wrought into various forms, at first by the immediate hand of the Creator, and after condensation wrought into various forms, at first by the immediate hand of the Creator, and ever since by the power of Nature, wch by vertue of the command Increase & Multiply, became a complete Imitator of the copies sett her by the Protoplast. Thus perhaps may all things be originated from aether.[71]

For nature is a perpetuall circulatory worker, generating fluids out of solids, and solids out of fluids, fixed things out of volatile, & volatile out of fixed, subtile out of gross, & gross out of subtile, Some things to ascend & make the upper terrestiall juices, Rivers and the Atmosphere; & by consequence others to descend for a Requitall to the former. And as the Earth, so perhaps may the Sun imbibe this Spirit copiously to conserve

---

[69] *Of Natures Obvious Laws*, NP; quoted Westfall (1980, 306); Dobbs (1982, 521).
[70] *Hyp., Corr.* I, 364.
[71] Letter to Oldenburg, 26 January 1676; *Hyp., Corr.* I, 414.

his Shineing, & keep the Planets from recedeing further from him. And they that will, may also suppose, that this Spirit affords or carryes with it thither the solary fewell and materiall Principle of Light; And that the vast æthereall Spaces between us, & the stars are for a sufficient repository for this food of the Sunn & Planets.[72]

The aether suggested a way of unifying physics, in fact, the only way, the aether being the only thing which could connect everything in a universal continuum. In this, it was not massively different from ideas like 'vacuum' and 'quantum fields', which provide exactly such a connection in quantum physics, or the unified field incorporated in the space–time manifold in Einstein's attempted extension of general relativity to a unified field theory. There was a problem, however, which Newton could not fail to notice as, within a few years, he developed his action-at-a-distance dynamics. The heavy, material aether of the Cartesian vortices seemed to be as incompatible with planetary dynamics as the vortices themselves. The 'æthereall Medium' he proposed in the *Hypothesis* was already diverging from the Cartesian aether. It was imagined to be 'much of the same constitution with air, but far rarer, subtiler & more strongly Elastic.'[73]

At this time, Newton thought that the fact that a pendulum motion decayed almost as quickly in a vessel exhausted of air by an air pump as it did in the open air, an experiment he had performed in the 1660s while he was at work on the *Quaestiones*, was a strong argument for the existence of an aether. He writes in the *Hypothesis*: 'Of the existence of this Medium the motion of a Pendulum in a glasse exhausted of Air almost as quickly as in the open Air, is no inconsiderable argument.'[74] However, as he came to think more and more strongly that the Cartesian vortices were incompatible with the laws of planetary motion, so he came to doubt the existence of the Cartesian aether. Eventually, he performed new experiments with a pendulum, with a box rather than a bob, which he took as proving that this aether did not exist. The resistance when the box was empty to when the box was full with metal was only in the proportion 77 to 78, suggesting

---

[72] *Hyp., Corr.* I, 366. The aethereal circulation described here may be compared with a real biological circulation driven by the light of the Sun (transpiration) which Newton prefigured in his student notebook. In a passage with the title 'Vegetables' (QQP, Beerling, 2015), he describes how solar energy is absorbed in plant shoots when their water particles are struck by light 'globules'. The heat generated causes fluid from the shoots to rise as sap through pores in the stem and leaves, where (as his diagram suggests) the water is evaporated into the atmosphere. Growth of the plant ceases when the pores are too narrow.

[73] *Hyp., Corr.* I, 364.

[74] *Hyp., Corr.* I, 364.

that the presence of matter made virtually none of the difference that would have been expected if a Cartesian aether was present.[75]

8.5. Gravity and the Aether

Having abandoned any trace of Descartes' material aether, Newton was unable to explain gravity — 'I feign no hypotheses'.[76] '...the cause of gravity is what I do not pretend to know, and therefore would take more time to consider of it.'[77] He could not identify the agent, if any, though he never completely gave up on an aethereal explanation. Writing to Richard Bentley on 25 February 1693, he argued against gravity being an innate property of matter:

> Tis unconceivable that inanimate brute matter should (without ye mediation of something else wch is not material) operate upon & affect other matter wthout mutual contact; as it must if gravitation in the sense of Epicurus be essential & inherent in it. And this is one reason why I desired you would not ascribe innate gravity to me. That gravity should be innate inherent & essential to matter so y^t one body may act upon another at a distance through a vacuum wthout the medium of anything else by & through wch their action or force may be conveyed from one to another is for me so great an absurdity that I beleive no man who has in philosophical matters any competent faculty of thinking can ever fall into it. Gravity must be caused by an agent acting constantly according to certain laws, but whether this agent be material or immaterial is a question I have left to ye consideration of my readers.[78]

There had to be an agent or active principle to connect bodies separated from each other acting across space: 'For two planets distant from one another by a long and empty interval will not approach one other by any force of gravity, nor will they act upon one another, except by the mediation of some active principle which comes between them both, and by means of which force is propagated from each into the other...'[79] He makes clear in a later draft Query that: 'By a vacuum I do not mean a space void of all

---

[75] Reported from memory in *Principia*, II, Section VI, General Scholium at the end following Proposition 31. The experiments were the first in a long line of attempts over the next few centuries aimed at proving that the aether either did or did not exist.

[76] *Principia*, General Scholium, 1713, 547, with the usual amendment of 'frame' to 'feign'.

[77] Newton to Bentley, 17 January 1693, *Corr.* III, 240.

[78] *Corr.* III, 253–254.

[79] Manuscript related to Classical Scholia for *Principia*; McGuire (1968, 196).

substances. Glass cannot attract light without a Medium. I mean only such a Vacuum as may be made by drawing Aer out of a vessel of glass.'[80]

If there was an aether, however, it had to be effectively 'dematerialised'. He wrote to Leibniz, saying that, while the heavens could not be full of a material aether, it was conceivable that some 'subtle matter', not disturbing the motions of the heavenly bodies, could be responsible:

> ...since all phenomena of the heavens and of the sea follow precisely, so far as I am aware, from nothing but gravity acting in accordance with the laws described by me; and since nature is very simple, I have myself concluded that all other causes are to be rejected and that the heavens are to be stripped as far as may be of all matter, lest the motions of planets and comets be hindered or rendered irregular. But if, meanwhile, someone explains gravity along with all its laws by the action of some subtle matter, and shows that the motion of planets and comets will not be disturbed by this matter, I shall be far from objecting.[81]

Newton's early theories of gravity, to which he returned at intervals, had involved a stream of aether particles of density ρ producing pressure ρv^2 by an impact, and his follower, Fatio de Duillier, had even developed a theory in which aether particles moved with random straight line motion until they were deflected by collisions with other particles.[82] However, Newton argued, in the General Scholium, of both 1713 and 1726 editions of the *Principia*, against gravity having a mechanical cause. The magnitude of a mechanical interaction would be proportional to the number of impacts per unit time, and so proportional to the surface area exposed. Gravity, however, is proportional to the mass of a body, and for a body of uniform density, to the volume, and not the surface area. This is interestingly reminiscent of the modern argument which says that quantum gravity fails to preserve locality, in that the entropy of black holes is proportional to the area of the event horizon, whereas locality requires the maximum entropy in any space to be decided by the volume. Interestingly, an aether force proportional to the surface area has connections with the holographic principle, also related to the entropy of black holes, in which the surface area is the sole source of information about a system. In addition, of course, the area parameter is a key aspect of gravity, if we are considering the gravitational influence of a massive object *outside* its own surface.

---

[80] Draft Q for *Opticks*, UL Cambridge, MS Add. 3970.3, NP.
[81] Newton to Leibniz, 16 October 1693, *Corr.* III, 287.
[82] Fatio (1688).

Newton also argued against gravity being 'innate'. Anxious to avoid any charge of atheism arising from his theory, he wrote to Bentley in 1693: 'You sometimes speak of gravity as essential & inherent to matter; pray do not ascribe that notion to me, for ye cause of gravity is what I do not pretend to know, & therefore would take more time to consider of it.'[83] And in another letter he stated, as we have seen, categorically that the idea 'That gravity should be innate, inherent & essential to matter' so as to act between bodies at a distance through a vacuum was such a manifest absurdity that no serious philosopher could fall into it.[84] Yet, he also stated at the same time that the cause of gravity could be either material or immaterial. There may have been occasions when an extreme situation would force out an idea which was latent in his thoughts but which ran counter to those which he supported officially.

In his preface to the second edition of the *Principia* (1713), Cotes wrote as follows:

> Since, then, all bodies, whether upon Earth or in the Heavens, are heavy, so far as we can make any experiments or observations concerning them; we must certainly allow that gravity is found in all bodies universally. And in like manner as we ought not to suppose that any bodies can be otherwise than extended, moveable, or impenetrable, so we ought not to conceive that any bodies can be otherwise than heavy.[85]

At almost exactly the same time, in response to the attacks of Leibniz on his philosophy, Newton wrote to the editor of the *Memoirs of Literature* in exactly the same vein, claiming that gravity was on the same footing with respect to bodies as their other unexplained properties: hardness, extension, durability, mobility and inertia.[86] In the same letter, he imagined an aether entirely without *vis inertiae* or mass, and acting entirely according to non-mechanical laws, but denied that it was either a miracle or an occult quality.

It is generally believed that both this letter and Cotes's preface misrepresented Newton's true position, and the earlier or late speculations on possible mechanical causes for gravity have sometimes been cited as supporting evidence. It is just possible, however, this was the only time when Newton actually did state his true position, the only time when he allowed his strong feelings to overcome the extreme caution with which he treated his public statements on the subject. He could not afford to have his system

---

[83] Newton to Bentley, 17 January 1693, *Corr.* III, 253.

[84] Newton to Bentley, 25 February 1693, *Corr.* III, 253–254.

[85] Cotes (1713).

[86] Newton to the editor of the *Memoirs of Literature*, *Corr.* V, 300.

condemned for atheism, and the idea would have been anathema to him in any case, so he invariably laid stress in his public writings on the divine origin of all physical phenomena. 'Innate' gravity was publicly associated with atheism, so he publicly rejected it. But, since, for him, everything took place in the sensorium of a divine being, no atheism could be attached to the fact that gravity had no material cause.[87]

He did not, of course, rule out the possibility of a material cause and speculated on a number of occasions on what such a cause could be, but it is likely that his usual position, at least at this period in his life, was that the cause of gravity was immaterial (the direct action of God) and that *in this sense* it was 'innate to matter', though the association with atheism made it necessary to reject such terminology. If gravity were a fundamental property of matter, the cause would not be mechanical; if it were not, then the cause would be explained mechanically; in believing it to be due to the direct action of God, he effectively required it to be non-mechanical. Cotes translated this idea into admirable pragmatism: 'For causes use to proceed in a continued chain from those that are more compounded to those that are more simple; when we are arrived at the most simple cause we can go no farther. Therefore no mechanical account or explanation of the most simple cause is to be expected or given; for if it could be given, the cause were not the most simple.'[88] Many such creative tensions occurred in Newton's work.

Even without a mechanistic explanation involving impact forces, Newton was always looking for what we might describe today as a 'non-local' way of connecting gravitating objects. A force that was neither an innate property nor caused by external agencies left few options. The one that seemed most likely towards the end of his life was an effectively dematerialised form of aether. Newton frequently assumed in his writings that aether was, or acted like, a highly rarefied form of 'air', and so he would certainly have been aware that, mathematically, dynamic models of both air and aether produced the same law of forces (mv^2/r) as systems governed by a central inverse square law force. The reason for this is the apparent coincidence that systems

---

[87]I believe this argument to be compatible with John Henry's position (1994) that gravity in Newton's understanding was a property of matter such that matter could act at a distance but the power was an active one, whether material or immaterial, which must have been directly conferred by God at the creation, not one that was an *essential* property of inanimate bodies in the same way as extension. As (in Bentley's words in the eighth Boyle lecture) 'a constant Energy infused into Matter by the Author of all things,' it did not require further divine intervention, but operated according to the observed laws of Nature.

[88]Cotes (1713).

acting under inverse square law forces and steady state dynamic systems subject to effectively constant force produce the same numerical relation between potential energy and mv^2 or between force and mv^2/r. Of course, the 'coincidence' is really an expression of the fact that inverse square laws are a natural result of three-dimensional space, inverse-square-law forces between elements leading to constant force on integration. But, coincidence or not, it was a fact of which Newton was aware after his early calculations of centripetal force, and he must have chosen a static model for his derivation of Boyle's law relating pressure and density in air (in *Principia*, Book II, Proposition 23), rather than a dynamic one, because it was simple and, for mathematical purposes, there was no need to do otherwise.

The result was that Newton's model for the force of refraction became mathematically equivalent to his formula for the velocity of light in aether, in the same way as the equation of motion of a satellite under gravity is mathematically equivalent, under the right conditions, to the formula for the velocity of sound in a medium. Its mathematical equivalence to yet another expression, the formula for the velocity of waves on a vibrating string, may have been responsible for Newton's extraordinary assertion in the famous 'classical scholia' that the ancients had discovered the law of gravity.[89] With the further stipulation that particles in random motion may be treated as having independent velocities in each of three perpendicular directions, these become equivalent to the respective calculations for pressure in radiation and material gases, and, indeed, to Newton's calculations involving the pressures caused by optical and gravitational aethers. So, it is not surprising, therefore, that several of the early exponents of the kinetic theory of gases, for example, Le Sage, Herapath and Waterston,[90] were inspired by the possibility that this theory might explain gravity, and that Fatio de Duillier's gravitational aether theory had such similarity to the kinetic model of a gas.

There were strongly abstract qualities to Newton's later conceptions of aether, based on the necessity of explaining action at a distance. Long before the late set of optical queries, Newton had begun to consider the 'possibility that God might replace the [Cartesian] aether [he] had already learned to employ. The role he attributed to God was exactly the role he attributed to the aether — an invisible medium that moves particles of matter in the manners they are observed to move.'[91] Such an 'immaterial aether' was

---

[89] Royal Society, Gregory MS 247; some material published in Gregory (1702); McGuire and Rattansi (1966).
[90] Le Sage (1756, 1761, 1784, 1818); Herapath (1816, 1821); Waterston (1843, 1892).
[91] Westfall (1971, 341).

implied in the letters to Bentley, written in 1692–1693, and 'To Newton, it was the infinite omnipotent God, who by His infinity constitutes absolute space and by His omnipotence is actively present throughout it.'[92] And, from draft material for the queries, we learn that 'the universal presence of God performed all the functions that Newton ascribed to the aether in the "Hypothesis of Light". God was an incorporeal aether who could move bodies without resistance to them in turn.'[93]

8.6. A New Unification

As a result of the experiments conducted by the Royal Society Demonstrator, Francis Hauksbee, between 1703 and 1713, Newton seems to have revived his speculations concerning aethers and vital spirits, and he designed a purpose-built experiment on radiant heat to test his conclusions that there may be a different kind of aether, rather more aethereal or 'subtle' than the first. This was carried out by Hauksbee's successor, John T. Desaguliers, in 1716, seemingly with positive results.[94] The earlier experiments had suggested that there was a subtle vibrating electric and elastic spirit which explained all electromagnetic phenomena, light, heat, and chemical and biological processes, all, that is, except gravity and the operations of the 'most secret works of Nature,' which he had been unable to uncover. However, gravity, most definitely, was not caused by either a particulate agent or an aetherial vibration with material properties, and certainly not by the electric and elastic spirit. But this meant that there was a disunity in nature: two independent modes of action. Even if the electric spirit could be described as a medium or aether of a subtle or rare kind, there could not be two aethers. Though modern cosmology, interestingly, seems to have found itself requiring two 'aethers' — dark matter and dark energy — Newton had already rejected this possibility in Query 28.

The experiment, which Desaguliers carried out to demonstrate that the aether postulated by Newton really existed, is described in Query 18:

> If in two tall cylindrical Vessels of Glass inverted, two little Thermometers be suspended so as not to touch the Vessels, and the Air be drawn out of one of these Vessels, and these Vessels thus prepared be carried out of a cold place into a warm one; the Thermometer *in vacuo* will grow warm as

---

[92] *ibid.*, 396.

[93] *ibid.*, 397; Q 19.

[94] Journal Book of the Royal Society, 15 November 1716, quoted in Guerlac (1967, 50). Newton's MS account of the experiment is in Add. 3970, No. 9, fol. 623. The published version is in Q 18.

much, and almost as soon, as the Thermometer which is not *in vacuo*. And when the Vessels are carried back into the cold place, the Thermometer *in vacuo* will grow cold almost as soon as the other Thermometer. Is not the Heat of the warm Room convey'd through the *Vacuum* by the Vibrations of a much subtiler Medium than Air, which after the Air was drawn out remained in the *Vacuum*? And is not this Medium the same with that Medium by which Light is refracted and reflected, and by whose Vibrations Light communicates Heat to Bodies, and is put into Fits of easy Reflexion and easy Transmission? And do not the Vibrations of this Medium in hot Bodies contribute to the intenseness and duration of their Heat? And do not hot Bodies communicate their Heat to contiguous cold ones, by the Vibrations of this Medium propagated from them into the cold ones? And is not this Medium exceedingly more rare and subtile than the Air, and exceedingly more elastick and active? And doth it not readily pervade all Bodies? And is it not (by its elastick force) expanded through all the Heavens?

Desaguliers himself commented on this experiment much later, in a letter to Hans Sloane, President of the Royal Society after Newton (4 March 1731):

That there is a subtile Medium ever finer than Light which serves in the Reflection, Refraction and Inflexion of Light, may be deduced from Phenomena, as is evident to those who read Sir Isaac Newton's Optics with such attention and skill as thoroughly to understand them. But whether there be a subtile Medium call'd Aether which is endued with such Properties as to be the Cause of Gravity by a mechanical Impulse, is only propos'd by Sir Isaac in his last English Edition of his Optics by way of Query. And whether the Medium acting upon Light be the same as the Aether hinted at for the Cause of Gravity, is very cautiously insinuated by our incomparable Philosopher, who mentions an Experiment (which I made before the Society by his Order) of a Thermometer in which the Liquor was rais'd as high by Heat in Vacuo Boyleano as the Liquor of another equally graduated Thermometer in Pleno, tho' the heat was not communicated to the first quite so fast; and thence deduces that there must be a Medium to convey the Heat to the Thermometer in the Glass where it was suspended, which Medium or Fluid must penetrate the Glass which is impervious to the Air....[95]

Commentators have sometimes been surprisingly negative about Newton's reaction to this experiment, but it is at least arguable that he had one of his great inspirations, when, at the last minute, he inserted four new Queries into his *Opticks*, then almost already in press. In this very last work

---

[95] Desaguliers (1731); Guerlac (1967, 51).

as an independent scientist, inspired by one of his old ideas, he guessed that there was just one medium extending throughout the heavens, one that was extremely rare and subtle, not made of ordinary matter, and that this medium acted in different ways, by its density causing gravity and by its elasticity allowing the vibrations which transmitted the effects of the electric force. There would only be one agent moving in one medium. Draft material for the Queries suggests that this aether, which had to be able to explain action at a distance, was a very subtle, and, in many ways, strongly abstract concept, perhaps nonmaterial, like the modern quantum vacuum: 'Qu. 17 Is there not something diffused through all space in & through wch bodies move without resistance & by means of wch they act upon one another at a distance in harmonical proportions of their distances.'[96] He had already written in a draft for the *Principia*:

> Vapours and exhalations on account of their rarity lose almost all percep-tible resistance, and in the common acceptance often lose even the name of bodies and are called spirits. And yet they can be called bodies in so far as they are the effluvia of bodies and have a resistance proportional to density. But if the effluvia of bodies were to change thus in respect of their forms so that they were to lose all power of resisting, and cease to be numbered among the phenomena, these I would no longer call bodies...[97]

Also, if an aether was composed of material structures, these, like all material structures, would have to gravitate, and this, as he argued in a manuscript of the 1690s, would very quickly cause the gravitational effect to blow up to an infinite value.

The new aether allowed a kind of unification of gravity and electricity, which was connected with the optical/gravity analogy he had already pursued in draft calculations, where the optical and gravitational forces showed the same variation in aether density gradient, and bodies acted on light at a distance in both cases to bend the rays. In the published Query 19, he wrote in explanation of total internal reflection:

> Doth not the Refraction of Light proceed from the different density of this Aethereal Medium in different places, the light receding always from the denser parts of the Medium? And is not the density thereof greater in free and open Spaces void of Air and other grosser Bodies, than within the Pores of Water, Glass, Crystal, Gems, and other compact Bodies? For when Light

---

[96] Draft Q for the *Opticks*, Add. 3970.3, f. 234[v], NP.
[97] Draft for *Principia*; McGuire (1966, 219).

passes through Glass or Crystal, and falling very obliquely upon the farther Surface thereof is totally internally reflected, the total Reflexion ought to proceed rather from the density and vigour of the Medium without and beyond the Glass, than from the rarity and weakness thereof.

The same aether was invoked in Query 20 as the possible cause of optical refraction, which could be treated mathematically as an analogous process to the deviation of a massive body from a straight line path under the action of a gravitational centripetal force, and was so in several manuscript calculations:

Doth not this Æthereal Medium in passing out of Water, Glass, Crystal, and other compact and dense Bodies into empty Spaces, grow denser and denser by degrees, and by that means refract the Rays of Light not in a point, but by bending them gradually in curve Lines? And doth not the gradual condensation of this Medium extend to some distance from the Bodies, and thereby cause the Inflexions of the Rays of Light, which pass by the edges of dense Bodies, at some distance from the Bodies?

The draft queries leading up to this had shown that the force on a ray of light in refraction was above 10^{26} times greater in proportion to the matter in them then the gravity of Earth towards the Sun in proportion to the matter in it. Since (as previously discussed) the Sun–Earth distance is 10^{18} times greater than the range of refractive forces, the gradient over the refractive range would be 10^{44}.

In Query 21, he spoke of the effect of a varying density gradient:

Is not this Medium much rarer within the dense Bodies of the Sun, Stars, Planets and Comets, than in the empty celestial Spaces between them? And in passing from them to great distances, doth it not grow denser and denser perpetually, and thereby cause the gravity of those great Bodies towards one another, and of their parts towards the Bodies; every Body endeavouring to go from the denser parts of the Medium towards the rarer? For if this Medium be rarer within the Sun's Body than at its Surface, and rarer there than at the hundredth part of an Inch from its Body, and rarer there than at the fiftieth part of an Inch from its Body, and rarer there than at the Orb of *Saturn*; I see no reason why the Increase of density should stop any where, and not rather be continued through all distances from the Sun to *Saturn*, and beyond. And though this Increase of density may at great distances be exceeding slow, yet if the elastick force of this Medium be exceeding great, it may suffice to impel Bodies from the denser parts of the Medium towards the rarer, with all that power which we call Gravity. And that the elastic force of this medium is exceeding great, may be gathered from the swiftness of its vibrations.

Whittaker, the great historian of aether theories, wrote of this passage in his Introduction to the 1931 edition of the *Opticks*:

> It is evident that aether-density plays much the same part in this hypothesis that the dielectric constant does in the electromagnetic theory of light; and Newton's suggestion regarding gravity resembles the modern hypothesis of Wiechert and his followers that gravitational potential is an expression of what may be called the specific inductive capacity and permeability of the aether, these qualities being affected by the presence of gravitating bodies; and matter (assumed to be electrical in nature) being attracted to places of greater dielectric constant.[98]

Whittaker took the mathematical reasoning to be based on Newton's formula for the velocity of a wave motion in a medium of elasticity k and density ρ:

$$v = \sqrt{\frac{k}{\rho}},$$

which, despite Newton's preference for a corpuscular model of light, was applied by analogy to the optical case, or at least to the aethereal medium through which light and the waves associated with light were supposed to be transmitted. The analogy between gravity and optical refraction was, of course, very much part of the Newtonian tradition, informing the work of Newton himself and, later, of his 18th-century follower, Michell.[99]

This concept (like the Newtonian approach to perturbed gravitational orbits) is not significantly different from the Einsteinian approach of attributing gravitational action to curvature. It is always possible to express a physical field, including Newtonian gravity, in covariant form (using the covariant derivative), and Élie Cartan would do exactly this for Newtonian gravity in 1923 and 1924.[100] Einstein proclaimed on many occasions that the general theory of relativity was an aether theory, and, in the sense that vacuum is a kind of immaterial aether, so is quantum mechanics. Oliver Lodge was quick to pick up the comments by Einstein on such aether interpretations of his theory, and, in an article intended for radio engineers and amateur radio enthusiasts, claimed that Einstein, here, was acting in the tradition of Newton! '...the old view of Newton that it must be responsible for gravitation too, is readily confirmed by the most recent discoveries of

---

[98]Whittaker (1931/1952, lxxi–lxxii).
[99]Michell (1784).
[100]Cartan (1923, 1924).

Einstein, which have shown that a gravitational field modifies the ether in its neighbourhood sufficient to cause a perceptible effect on the velocity of light. So that an enormous mass like the Sun acts as a feebly converging lens.'[101]

Lodge may well have been the first to describe the gravitational lens concept in a published article. Almost as soon as the 1919 results on light-bending were announced, he had pursued the analogy with optical refraction, with a 'gravitational effect upon the ether's elastic or dielectric coefficient, employing the same factor as expressive of a refractive index.' Soon afterwards, he gave the expression for the refractive index due to a potential V as $(1+2V/c^2)$, which is equivalent to $(1+2GM/Rc^2)$, and claimed that it was due to a refractivity $\mu - 1$ or 'extra aether tension' $\mu^2 - 1$, in the same way as Newton had said that aethereal tension would have to vary to produce gravitation. The refractive index method is now considered a perfectly acceptable way of deriving gravitational light bending, and it can be calculated assuming either light waves or light corpuscles.[102]

Even the velocity of light is included through the elasticity of the aether, Newton arguing in Query 21 that the elasticity of the aether could be estimated from the squared ratio of the speed of light to the speed of sound. The elasticity of the aether in proportion to its density must, therefore, be more than 490 billion times the corresponding proportion in air, and the aether 4.9×10^{11} times more elastic than air in proportion to its density:

> ...the Vibrations or Pulses of this Medium, that they may cause the alternate Fits of easy Transmission and easy Reflexion, must be swifter than Light, and by consequence above 700000 times swifter than Sounds. And therefore the elastick force of this Medium, in proportion to its density, must be above 7000000 × 7000000 (that is, above 490000000000) times greater than the elastick force of the Air is in proportion to its density. For the Velocities of the Pulses of elastick Mediums are in a subduplicate *Ratio* of the Elasticities and the Rarities of the Mediums taken together.

In a draft Query, he says: 'Resistance & Gravity are here ascribed to different causes; [resistance to the density of the Medium & velocity of the body, gravity to the variation of elasticity].' 'And yet the elasticity for causing gravity may be exceeding great.'[103]

---

[101] Lodge (1924, 878).

[102] Lodge (1919a, 1919b, 1921).

[103] Draft Q, UL Cambridge, MS Add. 3970.3, 262^r, NP.

Newton stated clearly that he had no idea what kind of aether could have such a property, and what kind of particles would compose it:

> And so if any one should suppose that *Æther* (like our Air) may contain Particles which endeavour to recede from one another (for I do not know what this *Æther* is) and that its Particles are exceedingly smaller than those of Air, or even those of Light: The exceeding smallness of its Particles may contribute to the greatness of the force by which those Particles may recede from one another, and thereby make that Medium more rare and elastick than Air, and by consequence exceedingly less able to resist the motions of Projectiles, and exceedingly more able to press upon gross Bodies by endeavouring to expand it self.[104]

Of course, in the modern theory of the vacuum, we avoid this problem by making the particles virtual and therefore not material at all.

Query 22 stressed the 'inconsiderable' nature of the resistance of the aether:

> May not Planets and Comets, and all gross Bodies, perform their Motions more freely, and with less resistance in this Æthereal Medium than in any Fluid, which fills all Space adequately without leaving any Pores, and by consequence is much denser than Quick-silver or Gold? And may not its resistance be so small as to be inconsiderable? For instance; If this Æther (for so I will call it) should be supposed 700000 times more elastick than our Air, and above 700000 times more rare; its resistance would be above 600000000 times less than that of Water. And so small a resistance would scarce make any sensible alteration in the Motions of the Planets in ten thousand Years. If any one would ask how a Medium can be so rare, let him tell me how the Air, in the upper parts of the Atmosphere, can be above an hundred thousand thousand times rarer than Gold. Let him also tell me, how an electrick Body can by Friction emit an Exhalation so rare and subtile, and yet so potent, as by its Emission to cause no sensible Diminution of the weight of the electrick Body, and to be expanded through a Sphere, whose Diameter is above two Feet, and yet to be able to agitate and carry up Leaf Copper, or Leaf Gold, at the distance of above a Foot from the electrick Body? And how the Effluvia of a Magnet can be so rare and subtile, as to pass through a Plate of Glass without any Resistance or Diminution of their Force, and yet so potent as to turn a magnetick Needle beyond the Glass?

---

[104]Q 21.

Query 28, arguing against a material aether, was notably retained when the new queries were added:

> ...if the Heavens were as dense as Water, they would not have much less Resistance than Water; if as dense as Quick-silver, they would not have much less Resistance than Quick-silver; if absolutely dense, or full of Matter without any *Vacuum*, let the Matter be never so subtil and fluid, they would have a greater Resistance than Quick-silver. A solid Globe in such a Medium would lose above half its Motion in moving three times the length of its Diameter, and a Globe not solid (such as are the Planets,) would be retarded sooner. And therefore to make way for the regular and lasting Motions of the Planets and Comets, it's necessary to empty the Heavens of all Matter, except perhaps some very thin Vapours, Streams, or Effluvia, arising from the Atmospheres of the Earth, Planets, and Comets, and from such an exceedingly rare Æthereal Medium as we described above.

The queries, as they were finally constituted, retained both of Newton's ways of describing the fundamental interactions in nature: forces between the material particles, and forces between the aetherial particles, which in turn acted on the material particles — in modern terms, we could describe these as local and non-local interactions, which my own work has shown operating in two spaces, real and vacuum.[105] In the Newtonian theory, the mutual repulsions of aether particles did not create problems over action at a distance because the repulsive force specifically manifested itself by filling the spaces between the aether particles.[106]

Of course, we no longer use the word 'aether', which has acquired too many non-scientific connotations and carries too much history, but we continue to use the word 'vacuum' without fully knowing what it really means. We regard all particles as quanta of continuous fields extending throughout space, and we say that charged particles 'polarize the vacuum' adjacent to them. In another sense, also, and possibly even closer to Newtonian ideas, is the fact that space has a non-zero temperature resulting from a microwave background radiation with an energy density of 10^{-14} Jm^{-3}, and the isotropic nature of this radiation, in a different context, makes it at least a very close

---

[105] For example, Rowlands (2007, 321–323, 485). Iliffe (2004) argues that Newton's 'efforts to bring together active principles, aethers and voids in various works were not failures that resulted from his "confusion" but were bold attempts to meld together concepts or ontologies that belonged to distinct enquiries.' I would add that they are not failures in any sense because they used his method of abstraction to reveal a duality that still remains fundamental in Nature.

[106] Heiman and McGuire (1971, 242–243).

approximation to Newton's absolute inertial rest frame, making it possible to come close to measuring the Earth's absolute motion through space.

Newton's theory was the first of a new kind, the progenitor of Einstein's attempts at a unified field theory and of such contemporary ideas as superstrings and supersymmetry, but it remains an anomaly in his work, a late stab at a problem that could not then be solved, the best available fit to data, though perhaps not the final answer, a return to a theory tried out, in a different form, long before and discarded during the period of his greatest creativity, when he wrote the *Principia* and the *Opticks*. Theories which attempt to create a physical or model-dependent union of gravity and electromagnetism do not explain the source of the unity, which is likely to be more fundamental and abstract, though they may offer a limited way of studying its effects. Whittaker described many such theories. They remind one of the 19th century theories of light associated with such names as Cauchy, Kelvin and Stokes, which preceded Maxwell's wholly abstract electromagnetic theory.

Newton's theory uniting gravity and electromagnetism is in some ways like these pre-Maxwellian theories of light with their aethers based on particular 'concrete' models, but it is also more abstract, though perhaps lacking the more fully abstract quality of his earlier work in gravity and mechanics. Today, we are confronted with the problem of unifying the four fundamental interactions now known. It seems unlikely that this can be done by inventing some special kind of 'particle' or some special kind of 'space–time' or some special 'history of the universe.' It is more likely that it will be understood through extension and explanation of the abstract and general principle of gauge invariance and the simple but general group symmetries uniting particles and their interactions.

It is possible to argue, as some have done, that it was merely to pacify his mechanistic critics that Newton resurrected his early speculations on material aethers, but it is likely that the cause was deeper and reflected a true need to find an overall unity in nature. Though he could not proceed to a new level of abstraction, the theory he did finally evolve was still only semi-mechanical and was in its own way an important achievement, for it was the beginning of the quest for a unified theory of the forces of nature, one which is yet to be successfully resolved.

8.7. Conclusion

Modern physics is necessarily very different from the physics of the Newtonian era. Quantum mechanics, relativity and the Standard Model, have now arrived, not to mention electromagnetic theory, thermodynamics and condensed matter physics, and Newton would hardly recognise them in their modern forms. However, it is astonishing how much of the work that Newton did achieve is compatible with these new developments and provides a smoothly continuous transition to them, despite all the talk about 'revolutionary' developments in the 19th and 20th centuries. Revolution, however, does not have to signify an abrupt transition or discontinuity. Several real scientific revolutions in the 20th century arrived without fanfare and nobody saw they were happening until after they had happened. This was certainly true of quantum mechanics, microelectronics, the computer, and the worldwide web. In fact, it is the insistence on 'revolution' as an abrupt transition, rather than as the opening up of an unexpected new field of inquiry, that, in my view, has been responsible for the major fault-line in physics today.

Abrupt discontinuities are also antithetical to the Newtonian way of doing physics, which was to create an abstract, generic pattern, which would be almost guaranteed to hold in all circumstances, and then to fit a series of increasingly less generic patterns to deal with individual phenomena as they arose. In fact, Newton's generic patterns do hold. They are nearly always fundamentally true, even when he extends his work outside the fields where his expertise has always been recognised — mathematics, mechanics, gravity and optics — and it is for this reason that his theories remain compatible with the work that greatly extended and transformed his tentative beginnings in the areas that we would now call quantum theory, relativity, electromagnetic theory, thermodynamics and astrophysics.

The discovery of this procedure — we have called it the 'system' — was far from obvious. No one else realised that this was what had to be done to transform physics into the universal system of knowledge that it became, and no one else fully realised that this was what had been done. The reason why Newton achieved it was probably because his particular pattern of thinking — his 'method' in our terminology — led him inevitably in this direction. To make vast connections from many apparently unrelated separate pieces of information is normally a very difficult thing to do. It requires, I think, a quite separate mode of mental activity from regular reasoning, one in which the unconscious plays the major role and the 'eureka moment' becomes the standard mode of activity. Though a large number of people say that their best results occur when they are not directly thinking about a subject, Newton, I believe, had this facility to an extraordinary degree, as well as an exceptional power in 'ordinary' reasoning to capitalise on his successes in this mode.

Because it allows the creator access to ideas at a very fundamental level, and shows the multiple connections almost instantly, the results of such a method are much more likely to be valid than the laboured results derived from trial and error applied to a succession of model-dependent hypotheses, even when they seem to provide immediate explanations for particular problems. It is because of this that Newton's ideas are of special interest at a time when physicists are trying to access a deeply fundamental level to arrive at a unified view of the subject.

It is possible to see the Newton of the *Opticks* and associated works as a consummate experimenter who concentrated on experimental findings where he was unable to provide an extensive mathematically-based theory. There is, however, another aspect to this story. Newton's work in optics and related fields also incorporated a massive attempt at theorising involving many phenomena. Here, he used the method that would ultimately bear fruit in gravitational theory and the *Principia*, of searching for the most fundamental and abstract conceptions that would serve as a basis for a generic explanation that would be almost obvious when it was reached.

Rather than setting up a complicated mathematical theory that relied on hypotheses, however, sophisticated it might appear, he thought about fundamentals in mostly qualitative terms. The origins of a theory of phenomena such as electricity and optics that did not respond readily to mathematical treatment should be sought in the most simple and general principles that could be found. In most cases, he was able to find such principles, and, to a large extent, they were qualitative or expressed in a relatively simple mathematical form. Whether or not they were taken up by his successors, they remain close to those that we use today. Eventually, principles like these were used to create the major physics developments that followed in the 19th and 20th centuries.

Currently, physics is concerned with attempting an even more fundamental development — a single theory incorporating all the particles and interactions and the abstract parameters used to describe them. Many people think that such a theory would be a mathematically sophisticated combination of the Standard Model of particle physics with general relativity, or a combination of modifications of these. I would venture to say that I don't think Newton would have started there. An even more complicated combination of two unexplained highly sophisticated theories is not a theory of foundations in the Newtonian style. However, since consistency with Newtonian ideas seems to have been a prescription for successful physical theories up to the present time, it does not seem unreasonable to suppose that it may also have the power to remove the fault-lines which have troubled physicists since the early part of the 20th century. We can, in this sense, see the work as a template for the future development of physics at the deepest and most fundamental level. Newton's work has not only been the ultimate origin of much of the physics that we use today. It may still have a great deal to contribute to it.

Bibliography

Abbreviations

Corr.   Turnbull, H. W., Scott, J. F., Hall, A. R. and Tilling, L. *The Correspondence of Isaac Newton*, 7 vols., Cambridge University Press, 1959–1977.

CUL   Cambridge University Library.

Hyp.   *An Hypothesis Explaining the Properties of Light*, 1675, *Correspondence*, I, 361–386.

KCC   King's College, Cambridge.

MP   Whiteside, D. T. *Mathematical Papers*, 8 vols., Cambridge University Press, 1967–1981.

NP   Iliffe, Rob and Mandelbrote, Scott. *The Newton Project*, http://www. newtonproject.sussex.ac.uk/.

NPA   Newman, William R. *The Chymistry of Isaac Newton Project*, http:// webapp1.dlib.indiana.edu/newton/.

NPC   Snobelen, Stephen D. The Newton Project Canada, http://www.isaac newton.ca/Isaac Newton.

OP   Shapiro, Alan E. *The Optical Papers of Isaiac Newton*, vol. 1, Cambridge University Press, 1984.

QQP   *Quaestiones Quaedam Philosophicae*, NP, *Certain Philosophical Questions: Newton's Trinity Notebook*, c 1664, eds. J. E. McGuire and M. Tamny, Cambridge, 1983.

Q1–31   *Opticks*, Queries 1–31.

Unp.   Hall, A. Rupert and Hall, Marie Boas. *The Unpublished Scientific Papers of Isaac Newton*, Cambridge University Press, 1962.

Works cited

Aboites, V., 'Some Remarks about Newton's Demonstrations in Optics', *British Journal for the Philosophy of Science*, 53, 455–458, 2003.

Abraham, M., 'Zur Elektrodynamik beweater Körper', *Rendiconti del Circolo Matematico Palermo*, 28, 1–28, 1909.

Abraham, M., 'Sull'elettrodinamica di Minkowski', *Rendiconti del Circolo Matematico Palermo*, 30, 33–46, 1910.

Alhazen (Ibn al-Haytham), *Opticae Thesaurus Alhazen Arabis Libri Septem*, ed. Friedrich Risner, Basel, 1572.

Ango, P., *L'Optique Divisé en Trois Livres*, Paris, 1682.

Armitage, A., *Edmond*, Nelson, London, 1966.

Ashkin, A. and Dziedzic, J. M., 'Radiation Pressure on a Free Liquid Surface', *Phys. Rev. Lett.*, 30, 139–142, 1973.

Baily, F., *An Account of the Revd John Flamsteed, the first Astronomer Royal, compiled from his own Manuscripts, and other authentic documents, never before published*, London, 1835, photo-reprint, Dawsons, London, 1966.

Barnett, S. M., 'Resolution of the Abraham-Minkowski dilemma', *Physical Review Letters*, 104, 070401, 2010.

Barnett, S. M. and Loudon, Rodney, 'The enigma of Optical Momentum in a Medium', *Philosophical Transactions*, A, 368, no. 1914, 927–939, 2010.

Barrow, I., *Lectiones XVIII, Cantabrigiae in Scholis Publicis Habitae; in Quibus Opticorum Phenomenon Genuinae Rationes Investigantur, ac Exponuntur (Lectiones opticae) (Optical Lectures)*, London, 1669.

Bartholin, E., *Experimenta Crystalli Islandica Disdiaclastici, Quibus Mira & Insolita Refractio Detegitur*, Copenhagen, 1669.

Bechler, Z., 'Newton's Search for a Mechanistic Model of Colour Dispersion: A Suggested Interpretation', Archive for History of Exact Sciences, 11, 1–37, 1973.

Bechler, Z., 'Newton's Law of Forces which are Inversely as the Mass: A Suggested Interpretation of his Later Efforts to Normalise a Mechanistic Model of Optical Dispersion', *Centaurus*, 18, 184–222, 1974.

Beerling, D. J., 'Newton and the Ascent of Water in Plants', *Nature Plants*, 1, 1–3, 2015.

Berlinski, D., *Newton's Gift*, Duckworth, 2001.

Bernoulli, J. II, *Recueil des Pièces Qui Ont Remportés les Aix de l'Académie*, tome iii, 1752.

Berry, M. V., *Physics World*, 10, December 1997, 42.

Berry, M. V., 'Geometry of Phase and Polarization Singularities, Illustrated by Edge Diffraction and the Tides'; 'Exuberant Interference: Rainbows, Tides, Edges, (de)Coherence...', *Philosophical Transactions*, A, 1023–37, 2002.

Biot, J.-B., *Life of Sir Isaac Newton*, transl. Howard Elphinstone, *Lives of Eminent Persons*, The Library of Useful Knowledge, London, 1833, NP.

Biot, J.-B., 'Analyse des Tables de Réfraction Construites par Newton, avec l'indication des Procédés Numériques par Lesquels il a pu les Calculer', *Journal des Savants*, 735–754, December 1836.

Birch, T., *The History of the Royal Society of London*, 4 vols., London, 1756–1757.

Blank, B. E., 'T'he Calculus Wars', Review of Jason Socrates *Bardi, The Calculus Wars: Newton, Leibniz, and the Greatest Mathematical Clash of All Time, Basic Books, 2007*, Notices of the AMS, 56, no. 5, 602–610, May 2009.

Bohr, N., 'On the Constitution of Atoms and Molecules', *Philosophical Magazine* (6), 26, 1–25, July 1913.

Bohr, N., 'The Quantum Postulate and the Recent Development of Atomic Theory', *Nature*, 121, 580–590, 14 April 1928; 'Discussions with Einstein on

Epistemological Problems in Atomic Physics', in *Albert Einstein: Philosopher-Scientist*, ed. P. A. Schilpp, Cambridge University Press, 1949.

Boltzmann, L., *Annalen der Physik*, 22, 31, 291–294, 15 May 1884.

Bošković, R. J., *Dissertationes quinque ad dioptricam Pertinentes*, Vienna, 1767, Dissertation I, §§142–204, 72–91 (originally published in German in 1765).

Boursma, P. J., *Physical Aspects of Colour*, 2nd English edition, New York, 1971.

Boyer, C. B., 'Early Estimates of the Velocity of Light', *Isis*, 33 (1), 24, 1941.

Boyer, C. B., *A History of Mathematics*, John Wiley, Chichester, 1968.

Boyle, R., *Experiments and Considerations Touching Colours*, December 1663.

Boyle, R., *New Experiments and Observations Touching Cold, or An Experimental History of Cold Begun*, 1665.

Bradley, J., 'A letter from the Reverend Mr. James Bradley...to Dr. Edmund Halley...giving an account of a new discovered motion of the fix'd stars', *Philosophical Transaction of the Royal Society*, 35, 637–661, 1728.

Brewster, Sir D., *The Life of Sir Isaac Newton*, John Murray, London, 1831.

Brewster, Sir D., *BA Report*, 1, 308–322, 1832.

Brewster, Sir D., *Memoirs of the Life, Writings, and Discoveries of Isaac Newton*, 2 vols., Edinburgh Thomas Constable & Co., 1855, Johnson Reprint Corporation, New York, 1865.

Brouncker, W., *Musica Compendium*, 1653 (transl. Descartes).

Brown, R. C., *The Tangled Origins of the Leibnizian Calculus: A Case Study of a Mathematical Revolution*, World Scientific, 2012.

Buchwald, J. Z., *The Rise of the Wave Theory of Light*, Chicago and London, 1989.

Buchwald, J. Z. and Feingold, M., *Newton and the Origin of Civilization*, Princeton University Press, 2012.

Cantor, G. N., *Optics after Newton: Theories of Light in Britain and Ireland, 1704–1840*, Manchester, 1983.

Cartan, É., *Ann. Ecole Norm.*, 40, 325, 1923.

Cartan, É., *Ann. Ecole Norm.*, 41, 1, 1924.

Casimir, H. B. G., *Proceedings, Koninklijke Akademie van Wetenschappen te Amsterdam*, 51, 793, 1948.

Cassini, J., *Mémoires de l'Académie des Sciences, depuis 1666 jusqu'à 1699*, 8, reprinted Paris, 1730.

Cavendish, H., c 1784, in *The Scientific Papers of the Honourable Henry Cavendish, F. R. S., Volume II: Chemical and Dynamical*, ed. E. Thorpe, London, 1921, 433–437.

Chandrasekhar, S., *Newton's Principia for the Common Reader*, 1995, 201.

Chapman, A., *England's Leonardo: Robert Hooke and the Seventeenth-Century Scientific Revolution*, IoP, 2004.

Charleton, W., *Physiologia Epicuro-Gassendo-Charltoniana*, 1654.

Christianson, G. E., *In the Presence of the Creator Isaac Newton & his Times*, The Free Press, 1984.

Clairaut, A.-C., 'Sur les Explications Cartésiennes et Newtoniennes de la Réfraction de la Lumière', Acaémie Royale des Sciences (Paris), *Mémoires Pour 1739*, 259–275, 1741.

Clairaut, A.-C., *Philosophical Transactions*, 48, 776–780, 1754.

Clairaut, A.-C., 'Mémoire sur les moyeux de perfectionner les lunettes d'approche, par l'usage d'objectifs composes de plusieurs matières différemment réfringentes', *Mémoires de l'académie Royale des Sciences, 1756* (Paris, 1762), 380–437, esp. 408–420 (read 1761).

Clark, D. H. and Clark, S. H. P., *Newton's Tyranny: The Suppressed Scientific Discoveries of Stephen Gray and John Flamsteed*, W. H. Freeman, 2000.

Cohen, I. B., Preface to I. Newton, *Opticks*, 1931, reprinted Dover Publications, 1952 (based on the fourth edition, 1730).

Cohen, I. B., 'Hypotheses in Newton's Philosophy', *Physis*, 8, 163–164, 1966.

Cohen, I. B., *Introduction to Newton's 'Principia'*, Cambridge, Massachusetts, 1971.

Cohen, I. B., 'A Guide to Newton's *Principia*', in *Isaac Newton The Principia A New Translation*, eds. I. Bernard Cohen and Anne Whitman, University of California Press, Berkeley and Los Angeles and London, 1999, 1–370.

Cohen, I. B. and Westfall, R. S., (eds.), *Newton A Norton Critical Reader*, W. W. Norton & Company, 1995, General Introduction, xi–xii.

Cohen, I. B. and Whitman, A., *Isaac Newton The Principia A New Translation*, Berkeley and Los Angeles and London: University of California Press, 1999.

Cotes, R., preface to the second edition of the *Principia*, 1713, English translation in Motte (1729).

Courtivron, G., *Traité d'optique, où l'on Donne la Theorie de la lumière Dans le Systeme Newtonien*, Paris, 1752.

d'Alembert, J.-le-R., *Pouscules Mathématiques ou Mémoires sur Différences Sujest de Géométrie, de Méchanique, d'optique, d'astronomie, &c.*, 8 vols., Paris, 1761–1780, Mémoire XX, 3, 341–413, 1764, esp. 371–379.

de Broglie, L., 'Ondes et quanta', *Comptes Rendus Hebdomadaires des Séances de l'Académie des Sciences*, 177, 507–510, 1923a.

de Broglie, L., 'Quanta de Lumière, Diffraction et Interférences', *Comptes Rendus Hebdomadaires des Séances de l'Académie des Sciences*, 177, 548–550, 1923b.

de Broglie, L., 'Les Quanta, la théorie Cinétique des Gaz et le principe de Fermat', *Comptes Rendus Hebdomadaires des Séances de l'Académie des Sciences*, 177, 630–632, 1923.

de Broglie, L., *Philosophical Magazine*, 47, 446, 1924.

Delisle, J.-N., *Mémoired de l'Académie Royale des Sciences*, 1715, 166.

de Morgan, A., *Newton: His Friend: His Niece*, 1885, reprinted, London, 1968.

Dempsey, L., 'Written in the flesh: Isaac Newton on the mind–body relation', *Stud. Hist. Phil. Sci.*, 37, 420–441, 2006.

Derham, W., *Philosophical Transactions*, 26, 2–35, 1708–1709.

Desaguliers, J. T., *Physico-Mechanical Lectures*, 1717.

Desaguliers, J. T., *Philosophical Transactions*, 35, 1728, 621.

Desaguliers, J. T., letter to Hans Sloane, 4 March 1731.

Descartes, R., *Cogitationes privatae*, 1619–1621, in *Œuvres de Descartes*, Charles Adam and Paul Tannery, 13 vols., Paris, 1897–1913, 10, 242–243.

Descartes, R., to I. Beeckman, 22 August 1634, in *Oeuvres de Descartes*, eds. C. Adam and P. Tannery, J. Vrin, Paris, 1971, I, 307.

Descartes, R., *La Dioptrique*, appended to *Discours de la Methode Pour Bien Conduire sa Raison*, Leyden, 1637, *Oeuvres*, eds. Charles Adam and Paul Tannery, 12 vols. and supplement (Paris, 1897–1913), 6, 97–101.

Descartes, R., *Principia Philosophiae* (1644), translated by Valentine Rodger Miller and Reese P. Miller, D. Reidel, Dordrecht, 1983.

Dobbs, B. J. T., 'Newton's Alchemy and His Theory of Matter', *Isis*, 73, 511–528, 1982.

Drexler, M., Sobey, I. J. and Bracher, C., 'Fractal Characteristics of Newton's Method on Polynomials', Report No. 96/14, 1996.

Duarte, F. J., 'Newton, Prisms, and the 'Opticks' of Tunable Lasers', *Optics and Photonics News*, 11 (5), 24–28, 2000.

Eddington, A. S., *Space, Time and Gravitation*, Cambridge, 1920.

Einstein, A., *Annalen der Physik*, 17, 132–148, 9 June 1905 (dated 17 March, received 18 March); translated in D. ter Haar, *The Old Quantum Theory*, Oxford, 1967, 91–107; also by A. B. Arons and M. B. Peppard, *American Journal of Physics*, 33, 367–374, 1965; and H. Boorse and L. Motz, *The World of the Atom*, New York 1966, 1, 544.

Einstein, A., *Jahrbuch für Radioaktivität*, 4, 411–462, 1907, 440 ff.

Einstein, A., *Physikalische Zeitschrift*, 10, 817–825, 1909; translated Anna Beck, *The Collected Papers of Albert Einstein. Vol. 2, The Swiss years: writings, 1900–1909*, Princeton University Press, 1987.

Einstein, A., *Annalen der Physik*, 35, 898–908, 1911.

Einstein, A., Foreword to I. Newton, *Opticks*, 1931, reprinted Dover Publications, 1952 (based on the fourth edition, 1730), lix–lx.

Einstein, A., Podolsky, B. and Rosen, N., 'Can Quantum-Mechanical Description of Physical Reality be Considered Complete?' *Physical Review*, 47 (10), 777–780, 1935.

Elliott, R. W., 'Isaac Newton as a Phonetician', *Modern Language Review*, 49, 1–12, 1954.

Elliott, R. W., 'Isaac Newton's 'Of an Universal language', *Modern Language Review*, 52, 1–18, 1957.

Euclid, L., 'The Optics of Euclid', translated by Harry Edwin Burton, *Journal of the Optical Society of America*, 35, no. 5, 357–372, May 1945.

Euler, L., 'Reflexions de Mr. L. Euler sur Quelques Nouvelles Experiences Optiques, Communiquées à l'Académie des Sciences. par Mr. Wilson', *Acta Academiae Scientiarum Petropolitanae*, 1, 71–77, 1746, 210–214.

Euler, L., *Sur la Perfection des Verres Objectifs Des Lunettes*, Mem. de l'Acad. de Berlin, 1747 (1753 and 1754).

Euler, L., Considerations sur Les Nouvelles Lunettes d'angleterre de Mr. Dollond et Sur le Principe Qui en est le Fondement', *Mémoires de l'Académie des Sciences de Berlin*, 1762 (1769), 226–248, *Opera omnia*, ser. 3., 8, 102–121, esp., §§10–12, 106–107.

Evans, J. and Rosenquist, M., *American Journal of Physics*, 54, 876–883, 1986.

Evans, W. A. B., *Physics Education*, 22, 330–336, 1987.

Fabri, H., *Dialogi Physici*, 1669.

Fatio de Duillier, N., *De la Cause de la Pesanteur (On the Cause of Weight)*, read to the Royal Society, 1688, published in Bernard Gagnebin, 'De la cause de la pesanteur. Mémoire de Nicholas Fatio de Duillier présenté à la Royal Society le 26 février 1690', *Notes and Records of the Royal Society*, 6, 106–160, 1949.

Feingold, M., *The Newtonian Moment: Isaac Newton and the Making of Modern Culture*, Oxford University Press, USA, 2005.

Fermat, P. de, to La Chambre, August 1657 and 1 January 1662; *Oeuvres de Fermat*, ed. Paul Tannery and Charles Adam, 4 vols. (Paris, 1891–1912), 2, 354–359 and 457–463.

Fizeau, A. H., *Comptes Rendus*, 29, 90, 1849.

Flamsteed, J., *Philosophical Transactions*, 7, 5118 (letter dated 16 November 1672).

Force, J. E., 'The God of Abraham and Isaac (Newton)', in *The Books of Nature and Scripture*, eds. J. E. Force and R. H. Popkin, Kluwer Academic, Dordrecht, 1994, 179–200.

Force, J. E., "Children of the resurrection" and 'children of the dust': Confronting mortality and immortality with Newton and Hume', in *Everything Connects: In Conference with Richard H. Popkin: Essays in his Honor*, eds. J. E. Force and D. S. Katz, Brill, Leiden, 1999, 119–142.

Foucault, J. B. L., *Comptes Rendus*, 30, 551, 1850.

Foucault, J. B. L., *Comptes Rendus* 55, 501, 1862.

Franklin, B., to Cadwallader Colden, Philadelphia, 23 April 1752, read at the Royal Society 1756; *Benjamin Franklin's Experiments: a new edition of Franklin's Experiments and Observations on Electricity*, ed. I. Bernard Cohen, Cambridge, 1941, 325; *Papers of Benjamin Franklin*, ed. L. W. Labaree, New Haven, Connecticut, 1961, 4, 297–301.

Fresnel, A. J., *Oeuvres*, 1, 385–439, 1816 (unpublished, 2 versions), 394.

Fresnel, A. J., *Oeuvres*, 1, 609–653; *Annales de Chimie et de Physique*, 17, 102–112, 167–196, 312–316, 1821.

Fresnel, A. J., *Annales de Chimie et de Physique*, 23, 32, 113; *Oeuvres*, 2, 215–238 (dated May and June 1823).

Fresnel, A. J., *Œuvres Complètes d'Augin Fresnel*, eds. H de Sénarment, E. Verdet and L. Fresnel, 3 vols., Paris, 1866–1870, 2: 821.

Gabbey, A., 'Newton, Active Powers, and the Mechanical Philosophy', in *The Cambridge Companion to Newton*, eds. I. Bernard Cohen and George E. Smith, Cambridge University Press, 2002, 329–357.

Gagnebin, B., 'De la cause de la pesanteur. Memoire de Nicholas Fatio de Duillier presente a la Royal Society le 26 fevrier 1690', *Notes and Records of the Royal Society*, 6, 106–160, 1949.

Galileo, *Discorsi*, Leyden, 1638; translated by Henry Crew and Alfonso de Salvio as *Dialogues Concerning Two New Sciences*, Dover, New York, 1954.

Gjertsen, D., *The Newton Handbook*, Routledge & Kegan Paul, 1986.

Gleick, J., *Isaac Newton*, HarperCollins, London, 2003.

Goethe, J. W. von, *Beyträge zur Optik*, 1791–1792.

Goos, F. and Hänchen, H., 'Ein neuer und fundamentaler Versuch zur Totalreflexion', *Annals of Physics*, (436) 7–8, 333–346, 1947.

Gorham, G., 'How Newton Solved the Mind–Body Problem', *History of Philosophy Quarterly*, 28, no. 1, 21–44, January 2011.

Gorlin, G., 'Demystifying Newton: The Force Behind the Genius', *The Objective Standard*, 3, no. 4, winter 2008–2009.

Gough, W., 'Mixing Scalars and Vectors — An Elegant View of Physics', *European Journal of Physics*, 11, 326–323, 1990.

Gouk, P. M., 'The Harmonic Roots of Newtonian Science', in eds. John Fauvel, Raymond flood, Michael Shortland and Robin Wilson, *Let Newton Be! A New Perspective on his Life and Works*, Oxford University Press, 1988, 101–125.

Gouk, P. M., *Music, Science and Natural Magic in Seventeenth-Century England*, New Haven and London, 1999, 224–257.

Gouy, L. G., *Journal de Physique*, 5, 354–362, 1886.

Gray, S., letter dated 3 January 1708, in R. A. Chipman, 'An Unpublished Letter of Stephen Gray on Electrical Experiments, 1707–1708', *Isis*, 45, 1954, 33–40.

Gray, S., *Philosophical Transactions*, 31, 1720, 140–148.

Gregory, D., *Astronomiae Physicae et Geometricae Elementa*, Oxford, 1702.

Gregory, D., *Elements of Physical and Geometrical Astronomy*, 1715 (translation of Latin of 1702), second edition 1726, reprinted Johnson Reprint Corporation, New York, 1972.

Grimaldi, F., *Physico-mathesis de Lumine, Coloribus, et Iride*, Bologna, 1665.

Guerlac, H., 'Newton's Optical Aether: His Draft of a Proposed Addition to His *Opticks*', *Notes and Records of the Royal Society*, 22, No. 1/2, 45–57, September 1967.

Guerlac, H., 'Hauksbee, Francis', *Dictionary of Scientific Biography*, 9, 169–176, 1972.

Guicciardini, N., *Isaac Newton on Mathematical Certainty and Method*, The MIT Press, Cambridge, Mass. and London, 2009.

Hall, A. R., 'Sir Isaac Newton's Notebook 1661–1665', *Cambridge Historical Journal*, 9, 239–250, 1948.

Hall, A. R., *From Galileo to Newton*, Dover, 1981.

Hall, A. R., *Isaac Newton Adventurer in Thought*, Cambridge University Press, 1992.

Halley, E., 'An account of the course of the tides at Tonqueen in a letter from Mr. Francis Davenport, July 15 1678 with the theory of them, at the Barr of Tonqueen, by the learned Edmund Halley, Fellow of The Royal Society', *Philosophical Transactions*, 14, 677–688, 1684.

Halley, E., 'Monsieur Cassini, his New and Exact Tables for the Eclipses of the First Satellite of Jupiter, reduced to the Julian Stile and Meridian of London', *Philosophical Transactions*, 18 (214), 237–256, 1694.

Halley, E., 'Some Remarks on the Allowances to be Made in Astronomical Observations for the Refraction of the Air', *Philosophical Transactions*, 31, 169–172, 1721.

Hänsch, T. W., 'Repetitively Pulsed Tunable Dye Laser for High Resolution Spectroscopy', *Appl. Opt.*, 11, 895–898, 1972.

Harrison, D. M., *The Influence of Non-linearities on Oscillator Noise Performance*, PhD Thesis, September 1988, Leeds University.

Hauksbee, F., 'Several Experiments on the Mercurial Phosphorus, made Before the Royal Society, at Gresham-College', *Philosophical Transactions*, 24, 2129–2135, 1704–1705.

Hauksbee, F., *Philosophical Transactions*, 25, 2332–2335, 1706.

Hauksbee, F., *Philosophical Transactions*, 25, 2374, 1707.

Hauksbee, F., 'An Experiment touching Motion given Bodies included in a Glass, by the Approach of a Finger Near its Outside: with other Experiments on the Effluvia of Glass', *Philosophical Transactions*, 26, 1708–1709, 82–86.

Hauksbee, F., *Physico-Mechanical Experiments on Various Subjects*, London 1709, second edition, 1719.

Hawes, J. L., 'Newton's 'Two Electricities'', *Annals of Science*, 27, 95–103, 1971.

Heaviside, O., 'A Gravitational and Electromagnetic Analogy, Part I', *The Electrician*, 31, 281–282, 1893.

Heaviside, O., 'A Gravitational and Electromagnetic Analogy, Part II', *The Electrician*, 33, 359, 1893.

Heaviside, O., *Electromagnetic Theory*, London, 1894, vol. 1, Appendix, 455–465

Heilbron, J. L., *Elements of Early Modern Physics*, University of California Press, 1982.

Heiman, P. M. and McGuire, J. E., 'Newtonian Forces and Lockean Powers: Concepts of Matter in 18th Century Thought', *Historical Studies in the Physical Sciences*, 3, 233–306, 1971.

Heisenberg, W., *Zeitschrift für Physik*, 33, 879–893, 18 September 1925 (received 29 July).

Heisenberg, W., *Zeitschrift für Physik*, 43, 172–198, 1927.

Hendry, J., *Centaurus*, 23, 230–251, 1980.

Henry, J., "Pray do not ascribe that notion to me': God and Newton's Gravity", in *The Books of Nature and Scripture: Recent Essays on Natural Philosophy, Theology and Biblical Criticism in the Netherlands of Spinoza's Time and the British Isles of Newton's Time*, eds. James E. Force and Richard H. Popkin, Kluwer Academic Publishers, Dordrecht, 1994, 123–147.

Herapath, J., 'On the Physical Properties of Gases', *Annals of Philosophy*, Robert Baldwin, 1816, 56–60.

Herapath, J., 'On the Causes, Laws and Phenomena of Heat, Gases, Gravitation', *Annals of Philosophy*, Baldwin, Cradock, and Joy, 9, 273–293, 1821.

Herivel, J., *The Background to Newton's Principia*, Oxford: Clarendon Press, 1965, 257–289.

Herschel, W., *Philosophical Transactions*, 1800, 255 ff, 284 ff, 293 ff.

Hestenes, D., *Space–Time Algebras*, Gordon and Breach, 1966.

Hobbes, T., *Tractatus Opticus*; in *Universae Geometricae, Mixtaque Synopsis, et Bini Refractionum Demonstratum*, ed. M. Mersenne, Paris, 1644, 576–587.

Hobden, H. and Mervyn, H., *The Cosmic Elk*, eighth edition 2009.

Hobden, M. K. 'John Harrison, Balthazar van der Pol and the Non-Linear Oscillator', *Horological Journal*, July 1981, 16–18, November 1981, 16–19, February 1982, 26–28, April 1982, 15–18, June 1982, 14–15.

Hobden M. K. 'As 3 is to 2', *Horological Science News*, November 2011, NAWCC Chapter, 161.

Home, R. W., 'Newton on Electricity and the Aether', in *Contemporary Newtonian Research*, ed. Z. Bechler, D. Reidel Publishing Company, 1982, 191–213.

Hooke, R., *Micrographia*, London, 1665.

Hooke, R., *An Attempt to Prove the Motion of the Earth by Observations*, 1674, reprinted in *Lectiones Cutlerianae*, 1679; reproduced in R. T. Gunther, *Early Science in Oxford*, Oxford University, Oxford, 1930, VII, 1–28.

Hooke, R., *Lectures of Light*, January 1680, *Posthumous Works*, 1705, reprinted Frank Cass & Co. Ltd., London, second edition, 1974, 78.

Hooke, R., *Discourse of the Nature of Comets* (1680), *Posthumous Works*, 1705, reprinted, Frank Cass & Co. Ltd., London, second edition, 1974.

Horsley, S., *The Power of God, deduced from the Computable Instantaneous Productions of it in the Solar System*, London, 1767.

Horsley, S., *Philosophical Transactions*, 60, 417–440, 1770 (read 20 December).

Huygens, C., *De vi centrifuga*, 1673, in *Oeuvres Complètes de Christiaan Huygens*, The Hague, 1929, XVI, 255–301.

Huygens, C., to Rømer, 11 November 1677, *Oeuvres Complètes* (1888–1950), 8, 41.

Huygens, C., *Traité de la Lumière* (Leyden, 1690), in *Oeuvres Complètes Publiés par la Société Hollandaise des Sciences*, 22 vols. (The Hague, 1888–1950), 19, 457–537 (intended to be read in August 1678, read May–August 1679, published with additions 1690); translated as *Treatise on Light* by S. P. Thompson (Macmillan, London, 1912, second edition, Chicago, 1950), 35–39.

Huygens, C., *Oeuvres completes*, 1888–1950, 22, 268.

Hyland, G. J. and Rowlands, P. (eds.), *Herbert Fröhlich: A Physicist Ahead of His Time*, University of Liverpool, 2006, second edition, 2008.

Iliffe, R., 'Abstract Considerations: Disciplines and the Incoherence of Newton's Natural Philosophy', *Studies in the History and Philosophy of Science*, 35 (3). 427–454, 2004.

Janiak, A., *Newton as Philosopher*, Cambridge University Press, 2008.

Jones, R. V. and Leslie, B., 'The measurement of Optical Radiation Pressure in Dispersive Media', Proc. Roy. Soc., A, 360, 347–363, 1978.

Jones, R. V. and Richards, J. C. S., 'The Pressure of Radiation in a Reflecting Medium', *Proc. Roy. Soc.*, A, 221, 480–498, 1954.

Kepler, J., *Ad Vitellionem Paralipomena, quibus Astronomiae pars optica*, 1604.

King, P., *The Life of John Locke*, second edition, Henry Colburn and Richard Bentley, London, 1830.

Kittel, C., *Quantum Theory of Solids*, Wiley, New York, 1987.

Kochiras, H., 'By ye Divine Arm: God and Substance in *De Gravitatione*', *Religious Studies*, 49, issue 03, 327–356, September 2013.

Koyré, A. and Cohen, I. B., 'Newton's Electric and Elastic Spirit', *Isis*, 51, 337, 1960.

Laplace, P. S., *Traité de Mécanique Céleste*, vol. 4, Paris, 1805 (read 27 May), 231–276.

Laplace, P. S., 'Sur la vitesse du son dans l'air et dans l'eau', *Annales de Chimie*, 3, 238–241, 1816; *Oeuvres complètes*, vol. 14, 297–300.

Lehn, W. H., 'Isaac Newton and the Astronomical Refraction', *Applied Optics*, 47, no. 34, H95–105, 1 December 2008.

Le Sage, G.-L., 'Letter à une académicien de Dijon', *Mercure de France*, 1756 153–171.

Le Sage, G.-L., *Essai de Chymie Méchanique*, not published, private print, 1761.

Le Sage, G.-L., 'Lucrèce Newtonien', *Memoires de l'Academie Royale des Sciences et Belles Lettres de Berlin*, 404–432, 1784, translated as *The Newtonian Lucretius*, in Samuel P. Langley, 'The Le Sage Theory of Gravitation', *Annual Report of the Board of Regents of the Smithsonian Institution*, 139–160, 30 June 1898.

Le Sage, G.-L., 'Physique Mécanique des Georges-Louis Le Sage', in Pierre Prévost, *Deux Traites de Physique Mécanique*, J. J. Paschoud, Geneva & Paris, 1818, 1–186.

Levenson, T., *Newton and the Counterfeiter*, Faber and Faber, 2009.

Lewis, G. N., 'The Conservation of Photons', *Nature*, 118, 874, 18 December 1926.

Lindberg, D. C., *Roger Bacon and the Origins of Perspectiva in the Middle Ages: A Critical Edition and English Translation of Bacon's Perspectiva, with Introduction and Notes*, Oxford University Press, 1996, 143.

Lindberg, D. C., 'Late Thirteenth-Century Synthesis in Optics', in ed. Edward Grant, *A Source Book in Medieval Science*, Harvard University Press, 1974, 396.

Lodge, O., 'On the Question of Absolute Velocity, and on the Mechanical Function of an Æther, with some Remarks on the Pressure of Radiation', *Philosophical Magazine*, 5th series, 46, 414–426, October 1898.

Lodge, O., 'Discussion on the Theory of Relativity', *Monthly Notices of the Royal Astronomical Society*, 80, 96–118, 2 December 1919a.

Lodge, O., *Nature*, 104, 354, 4 December 1919b (dated 30 November).

Lodge, O., *Philosophical Magazine* (6), 41, 549–557, April 1921.

Lodge, O., 'The Ether: Modern Theories and their Applications', *Harmsworth's Wireless Encyclopaedia*, London, March 1924, 876–882.

Lohne, J. A., 'Isaac Newton: The Rise of a Scientist 1661–1671', *Notes and Records of the Royal Society*, 20, no. 2, 125–139, December 1965.

MacKay, R. H. and Oldford, R. W., 'Scientific Method, Statistical Method and the Speed of Light', *Statistical Science*, 15 (3), 254–278, 2000.

Maignan, E., *Perspectiva Optica*, Rome, 1648.

Malus, E., *Mémoires de Physique at de Chimie de la Société d'Arceuil*, 2, 143–158, 1809 (read 12 December 1808).

Manuel, F. E., *A Portrait of Isaac Newton*, Cambridge, Belknap Press of Harvard University Press, 1968, reprinted Da Capo Press, New York, 1990.

Maraldo, G. F., *Mémoired de l'Académie Royale des Sciences*, 1723, 111.

Marci, M., *Thaumantias*, Prague, 1648.

Maxwell, J. C., *Nature*, 8, 437–441, 1873; *Scientific Papers*, 2, 376.

McGuire, J. E., 'Body and void in Newton's *De Mundi Systemate*: Some New Sources', *Archive for History of Exact Sciences*, 3, 206–248, 1966.

McGuire, J. E., 'Transmutation and immutability: Newton's Doctrine of Physical Qualities', *Ambix*, 14, 69–95, 1967.

McGuire, J. E., 'Force, Active Principles, and Newton's Invisible Realm', *Ambix*, 15, 1968, 154–208.

McGuire, J. E. and Rattansi, P. M., 'Newton and the 'Pipes of Pan", *Notes and Records of the Royal Society* 21, 108–143, 1966.

McGuire, J. E. and Tamny, M., *Certain Philosophical Questions: Newton's Trinity Notebook*, Cambridge University Press, 466–489, 1983.

Melvill, T., *Philosophical Transactions*, 48, 261–70, 1753.

Mersenne, M., *De l'utilitie' de l'harmonie in Harmonie Universelle*, Paris, 1636.

Mersenne, M., *Cogita Physico-Mathematica*, Paris, 1644.

Michell, J., 'On the Means of Discovering the Distance, Magnitude, &C. of the Fixed Stars, in Consequence of the Diminution of the Velocity of Their Light, in Case Such a Diminution Should Be Found to Take Place in Any of Them, and Such Other Data Should Be Procured from Observations, As Would Be Farther Necessary for That Purpose', *Royal Society of London Philosophical Transactions*, 74, 35–57, 1784, read 27 November 1783.

Michels, W. C. and Patterson, A. L., *Physical Review*, 60, 587, 1941.

Minkowski, H., 'Die Grundaleihunaean für die elektromagnetischen Vorgänae in beweaten Körpen', *Nachr. Könial. Ges. Wiss. Göttingen*, 53–111, 1908; reprinted *Mathematische Annalen*, 68, 472–525, 1910.

Motte, A., *The Mathematical Principles of Natural Philosophy. By Sir Isaac Newton. Translated into English*, 2 vols., London, 1729.

Newman, W. R., 'The Background to Newton's Chymistry', in *The Cambridge Companion to Newton*, eds. I. Bernard Cohen and George E. Smith, Cambridge University Press, 2002, 358–369.

Newton, I., *Opticks*, first edition, 1704, Latin edition, 1706, second English edition, 1717, NP; fourth edition, 1730 (edition cited); *Opticks* (preface by I. B. Cohen, foreword by A. Einstein, introduction by E. T. Whittaker), Dover Publications, 1952 (based on the fourth edition, 1730).

Newton, I., *Principia*, first edition, 1687, second edition, 1713, third edition, 1726, NP; English translation by Andrew Motte, 1729 (edition cited); revised by Florian Cajori, 1934; I. Bernard Cohen and Anne Whitman, *Isaac Newton The Principia A New Translation*, University of California Press, Berkeley and Los Angeles and London, 1999.

North, R., *The Lives of the Hon. Francis North, Baron Guildford; the Hon. Sir Dudley North; and the Hon. And Rev. Dr John North, together with the Autobiography of the Author*, George Bell and Sons, London, 3 vols., ed. Augustus Jessopp, 1890.

Oldenburg, H., *The Correspondence of Henry Oldenburg.*, eds. A. R. Hall and M. B. Hall, 13 vols., London and Madison, WI, 1965–1986.

Pais, A., *Inward Bound. Of Matter and Forces in the Physical World*, Oxford, 1986.

Palter, R. and Hynd, J., 'Early Measurements of Magnetic Force', Isis, 63, no. 4, 544–558, December 1972.

Pask, C., *Magnificent Principia Exploring Isaac Newton's Masterpiece*, Prometheus Books, 2013.

Poisson, S. D., to Fresnel, 6 March 1823a, in Fresnel, *Oeuvres*, 2, 886–889.

Poisson, S. D., *Annales de Chimie et de Physique*, 9, 1823b.

Poisson, S. D., *Annales de Chimie et de Physique*, 22, 250 (dated 24 March 1823c) in Fresnel, *Oeuvres*, 2, 192–205.

Poisson, S. D., note of 29 June 1823d, appended to A. J. Fresnel, *Annales de Chimie et de Physique*, 23, 32, 113, in Fresnel, *Oeuvres*, 2, 226–227.

Poynting, J. H., *Philosophical Magazine*, 9, 169–171, 1905a (read to Section A of the BA at Cambridge, August 1904); 393–400, April 1905 (Presidential Address to the Physical Society, 10 February).

Poynting, J. H., *Philosophical Magazine*, 9, 393–400, April 1905b.

Poynting, J. H. and Barlow, G., *BA Report*, 385, 1909.

Poynting, J. H. and Barlow, G., *Proc. Roy. Soc.*, A 83, 534–546, 1910.

Preston, T., *The Theory of Light*, second edition, 1895.

Refsdal, S. and Surdej, J., 'Gravitational Lenses', *Reports Progress in Physics*, 57, 117, 1994.

Roberval, G. P. de, *De Longitudine Trochoidis Proposition*, c 1640, first published in *Divers Ouvrages*, 1693, *Mémoires de l'Académie des Sciences*, 6, 419–423.

Rømer, O., 1676, 'Démonstration Touchant le Mouvement de la Lumière Trouvé par M. Römer de l'Academie Royale des Sciences', *Journal des Sçavans*, 7 December 1676, 223–236.

Rømer, O., 'A Demonstration concerning the Motion of Light', *Philosophical Transactions*, 12, no. 136, 25 June 1677, 893–894.

Ronchi, V., *The Nature of Light: An Historical Survey*, translated by V. Barocas, London, 1970 (originally published in Italian as *Storia della Luce*, 1939), 130.

Rosenfeld, L., *Bohr, On the Constitution of Atoms and Molecules*: Papers of 1913 reprinted from the *Philosophical Magazine* with an Introduction by L. Rosenfeld, Copenhagen and New York, 1963, Introduction, xxxix–xl.

Rowlands, P., *Newton and the Concept of Mass–Energy*, Liverpool University Press, 1990, ISBN 0 85323 187 7.

Rowlands, P., *Waves Versus Corpuscles: The Revolution That Never Was*, PD Publications, Liverpool, 1992, ISBN 1 873694 01 6.

Rowlands, P., *A Revolution Too Far: The Establishment of General Relativity*, PD Publications, Liverpool, 1994, ISBN 1 873694 03 2.

Rowlands, P., 'Theology and Modern Physics', *IoP History of Physics Group Newsletter*, 13, 57–62, 2000.

Rowlands, P., 'Isaac Newton: The Gifted Genius', *Physics World*, 14, 55, October 2001 (review of *Newton's Gift*, by David Berlinski).

Rowlands, P., 'Sir Oliver Lodge', *Dictionary of National Biography*, 34, 279–282, Oxford University Press, 2004.

Rowlands, P., 'The Difficult One', *New Scientist*, 187, no. 2513, 34–35, 20 August 2005.

Rowlands, P., *Zero to Infinity The Foundations of Physics*, World Scientific, 2007.

Rowlands, P., *The Foundations of Physical Law*, World Scientific, 2014a.

Rowlands, P., *How Schrödinger's Cat Escaped the Box*, World Scientific, 2014b.

Rowlands, P. and Attwood, T. V. (eds.), *War and Peace: The Life and Work of Sir Joseph Rotblat*, University of Liverpool, 2006.

Russell, B., 'The Metaphysician's Nightmare', in *Nightmares of Eminent Persons*, George Allen and Unwin Ltd, 1954.

Rutherford, E., *Manchester Literary and Philosophical Society Proceedings*, 55, 18–20, 1911a (abstract of paper read on 7 March).

Rutherford, E., *Philosophical Magazine* (6), 21, 669–688, May 1911b (dated April).

Rutherford, E., to Niels Bohr, 20 March 1913, in Bohr (1981), 112.

Sabra, A. I., *Theories of Light from Descartes to Newton*, London, 1967.

Sakellariadis, S., 'Descartes' Experimental Proof of the Infinite Velocity of Light and Huygens' Rejoinder', *Archive for History of the Exact Sciences*, 26, no. 1, 1–12, 1982.

Sarton, G., *Ancient Science Through the Golden Age of Greece*, Courier Dover, 1993.

Sauveur, J., 'Sur la détermination d'un son fixe', *Histoires de l'Académie Royale des Sciences*, Paris, 1700, 166–178.

Schrödinger, E., 'Die gegenwärtige Situation in der Quantenmechanik (The present situation in quantum mechanics)', *Naturwissenschaften*, 23, 807, November 1935, translated in *Quantum Theory and Measurement*, eds. J. A. Wheeler and W. H. Zurek, Princeton University Press, 1983.

Shapiro, A. E., *Archive for History of Exact Sciences*, 11, 134–266, 1974.

Shapiro, A. E., 'Newton's 'Achromatic' Dispersion Law: Theoretical Background and Experimental Evidence', *Archive for History of Exact Sciences*, 21, 91–128, 1979, 111.

Shapiro, A. E., 'Huygens' 'Traité de la Lumière' and Newton's 'Opticks': Pursuing and Eschewing Hypotheses', *Notes and Records of the Royal Society*, 43, 223–247, 1989.

Shapiro, A. E., 'The Optical Lectures and the Foundations of the Theory of Optical Imagery', in Chapter 2 of *Before Newton: The Life and Times of Isaac Barrow*, ed. Mordecai Feingold, Cambridge University Press, 1990, 105–178.

Shapiro, A. E., *Fits, Passions, and Paroxysms*, Cambridge University Press, 1993.

Shapiro, A. E., 'Newton's optics and atomism', in *The Cambridge Companion to Newton*, eds. I. Bernard Cohen and George E. Smith, Cambridge University Press, 2002, 227–255.

Shapiro, A. E., 'Newton's Experimental Philosophy', *Early Science and Medicine*, 9, 185–217, 2004.

Shapiro, A. E., 'Skating on the edge: Newton's Investigation of Chromatic Dispersion and Achromatic Prisms and Lenses', in *Wrong for the Right Reasons*, eds. J. Z. Buchwald and A. Franklin, Springer, 2005, 99–125.

Shapiro, A. E., 'Newton's Optics', in *The Oxford Handbook of the History of Physics*, eds. Jed Buchwald and Robert Fox, Clarendon Press, Oxford, 2013, 166–198.

Silverman, M. P., Review of *David H. Clark and Stephen P. H. Clark, Newton's Tyranny: The Suppressed Scientific Discoveries of Stephen Gray and John Flamsteed, American Journal of Physics*, 71, no. 5, 508, May 2003.

Smith, G. E., 'Essay Review: *Newton's Principia for the Common Reader*, by S. Chandrasekhar', *Journal for the History of Astronomy*, 17, 353–361, 1996.

Smith, R., *Compleat System of Opticks*, 1738.

Smolinski, R., 'The Logic of Millennial Thought: Sir Isaac Newton among his Contemporaries', in *Newton and Religion: Context, Nature, and Influence*, eds. J. E. Force and R. H. Popkin, Kluwer Academic, Dordrecht, 259–290.

Snobelen, S. D., 'Lust, Pride and Ambition, Isaac Newton and the Devil', in *Newton and Newtonianism: New Studies*, eds. James E. Force and Sarah Hutton, Dordrecht, 2004, 155–181.

Soldner, J. von, *Astronomisches Jahrbuch für das Jahr 1804*, Späthen, Berlin, 161, 1801, dated March 1801, reprinted by P. Lenard in *Annalen der Physik*, 65, 1921, 593, translated in S. L. Jaki, *Foundation of Physics*, 8, 927–950, 1978.

Spence, J., *Anecdotes*, 1820.

Spencer, J. H., *The Eternal Law: Ancient Greek Philosophy, Modern Physics, and Ultimate Reality*, Param Media, 2012.

Stachel, J. and Toretti, R., 'Einstein's First Derivation of Mass–Energy Equivalence', *American Journal of Physics*, 50, 760–763, 1982.

Stein, H., 'Newton's Metaphysics', in *The Cambridge Companion to Newton*, eds. I. Bernard Cohen and George E. Smith, Cambridge University Press, 2002, 256–307.

Stuewer, R. H., *Isis*, 60, 392–394, 1969.

Stuewer, R. H., *Isis*, 61, 188–205, 1970.

Stukeley, W., *Memoirs of Sir Isaac Newton's Life* (1752), ed. A. Hastings White, Taylor & Francis, London, 1936.

Stukeley, W., *The Family Memoirs of the Rev. William Stukeley, M. D. and the Antiquarian and Other Correspondence of William Stukeley, Roger & Samuel Gale, Etc*, Publications of the Surtees Society, 1882.

Tangherlini, F. R., *Nuovo Cimento Supplementary*, 20, 1, 1961.

Tangherlini, F. R., *Physical Review*, A 12, 139–147, 1975.

Taylor, B., *Methodus Incrementorum Directa & Inversa*, London, 1715.

Taylor, B., *Philosophical Transactions*, 29, no. 344, 294–295, June–August 1715 (letter to Hans Sloane, 1714).

Taylor, B., 'An Account of some Experiments relating to Magnetism', *Philosophical Transactions*, 31, no. 368, 204–208, 1720–1721 (May–August 1721) (letter to Hans Sloane, 1714, further extract).

Thomson, J. J., *Conduction of Electricity through Gases*, Cambridge, 1903; second edition, 1906.

Tressan, C. de, *Essai Sur le Fluide Électrique, Considéré Comme Agent Universel*, Paris 1786, written in the 1740s.

Truesdell, C., 'Reactions of Late baroque Mechanics to Success, Conjecture, Error and Failure in Newton's *Principia*', in Robert Palter, *The Annus Mirabilis of Sir Isaac Newton 1666–1966*, The M.I.T. Press, Cambridge, Massachusetts and London, 1967, 192–232.

Vavilov, S. I., *Lekcii po Optike*, translation of Newton's *Optical Lectures*, Moscow/Leningrad, 1946.

Voltaire, *Dictionnaire Philosophique*, 1757.

Waldman, G., *Introduction to Light: The Physics of Light, Vision, and Color*, Dover, 2002.

Walker, G. B., Lahoz, D. G., and Walker, G., 'Measurement of the Abraham force in a barium titanate specimen', *Canadian Journal of Physics*, 53, 2577–2586, 1975.

Walsh, D., Carswell, R. F. and Weymann, R. J., *Nature*, 279, 381, 1979.

Waterston, J. J., *Thoughts on the Mental Functions*, 1843, 'On the Physics of Media that are Composed of Free and Perfectly Elastic Molecules in a State of Motion', *Philosophical Transactions* A, vol. 183, 1–79, 1892 (rejected by the Royal Society in 1845).

Weaver, J. H., *The World of Physics*, 3 vols., New York, 1987, with additional notes by L. Motz and D. McAdoo, 1, 511.

Westfall, R. S., 'Short-writing and the State of Newton's Conscience', *Notes and Record of the Royal Society*, 18, 10–16, 1963.

Westfall, R. S., 'Isaac Newton's Coloured Circles Twixt Two Contiguous Glasses', *Archive for History of Exact Sciences*, 2, 181–196, 1965.

Westfall, R. S., *Force in Newton's Physics*, Macdonald, London and New York, 1971.

Westfall, R. S., 'Newton and the Fudge Factor', *Science*, 179, no. 4075, 751–758, 23 February 1973.

Westfall, R. S., 'The Role of Alchemy in Newton's Career', in *Reason, Experiment and Mysticism in the Scientific Revolution*, eds. M. L. Righini Bonelli and William R. Shea, New York: Science History Publications, 1975, 189–232.

Westfall, R. S., *Never at Rest*, Cambridge University Press, 1980.

Whiston, W., *Memoirs of the Life and Writings of Mr. William Whiston... Written by Himself*, London, 1749; second edition, 2 vols., London, 1753.

White, M., *The Last Sorcerer*, Fourth Estate, 1997.

Whiteside, D. T., *Centaurus*, 24, 288–315, 1980.

Whittaker, E. T., Introduction to I. Newton, *Opticks*, 1931, reprinted Dover Publications, 1952 (based on the fourth edition, 1730).

Whittaker, E. T., *A History of the Theories of Aether and Electricity*, 1st edition, London, 1910, 2nd edition, London and New York, 1951–1953, 1, 95–96.

Will, C. M., *American Journal of Physics*, 56, 413–415, 1988.

Wood, A., *Thomas Young Natural Philosopher 1773–1829*, Cambridge, 1954.

Woollard, E. M. and Clemence, G. M., *Spherical Astronomy*, New York, 1966.

Wróblewski, A., 'de Mora Luminis: A Spectacle in Two Acts with a Prologue and an Epilogue', *American Journal of Physics*, 53 (7), 620–630, July 1985.

Young, T., *Philosophical Transactions*, 90, 106–150, 1800.

Young, T., 'The Bakerian Lecture. On the Theory of Light and Colours', read 12 November 1801, *Philosophical Transactions*, 92, 12–48, 1802a.

Young, T., *Philosophical Transactions*, 92, 387–397, 1802b (read 1 July).

Young, T., *A Course of Lectures on Natural Philosophy and the Mechanical Arts*, 2 vols., London, 1807.

Young, T., to Arago, 12 January 1817a; T. Young, *Works*, 1, 383.

Young, T., 'Chromatics', in the *Supplement to the fourth, fifth and sixth editions of the Encyclopaedia Britannica*, 3, 141–163, 1824 (article written September–October 1817b); *Works*, 1, 279–342.

Index

A

A New Theory about Light and Colours, 33, 236–238
abduction, 7
aberration of light, 105
aberration, chromatic, 32, 231
aberration, spherical, 231
Abraham, M., 129
Abrégé de la chronologie, 54
absolute inertial rest frame, 265
absolute space, 35, 257
absolute time, 35
absolute velocity, 176
acceleration, 61, 90
acids, 176
action and reaction, 156, 175
action at a distance, 141, 181, 214, 219, 248, 251, 256, 259, 264
active powers, 214
active principles, 158, 194, 215, 250, 252, 264
aether, 1, 9–10, 12, 18, 23, 36, 44, 68, 81–82, 85, 89–91, 94, 103, 107, 116, 118, 127, 140–141, 149, 165, 167, 173–174, 176, 191, 193–195, 198, 202, 240, 244, 248–251, 253–257, 259–265
aethereal currents, 193
aethereal density gradient, 150, 173, 260
aethereal force, 253, 258, 260, 262–263
affinity, 18, 208
air, 69, 75, 89, 156, 174, 176, 194, 217, 250–251, 255, 258, 263
air resistance, 62, 64

alchemy, 8–10, 13, 17, 22, 33, 36, 40, 45, 49, 137, 156, 158, 194–195, 197, 209, 247, 249–250
algebra, 30, 35, 59, 63, 110, 135, 193
amber, 176, 208, 225
amplitude modulation, 71
An Hypothesis Explaining the Properties of Light, 17, 36, 58, 66, 87, 89–90, 96, 101–102, 126, 165, 174, 192–195, 197–198, 238–239, 249, 251, 257
analysis, 24
ancient wisdom, 17
annus mirabilis, 31
Ango, Pierre, 108
angular frequency, 127
angular momentum, 170
anomalous dispersion, 127
Anthony, St., 35
antimony, 43
Arbuthnot, John, 50, 53
Archimedes, 15, 43
Arius, 34–35
arsenic, 232
Ashkin, A., 185
astrophysics, 266
Athanasius, 34–35
atmospheric refraction, 75, 138, 145, 147
atomism, 29, 32, 179, 237
atoms, 18, 29, 186, 207, 237–238
Atterbury, Bishop Francis, 50
attractive forces, 181
Aubrey, John, 54

Printed in the United States
By Bookmasters